Mathematics

6th edition

for **Elementary** and **Middle School Teachers**

Sybilla Beckmann

University of Georgia

Pearson

* Available in print or for download.
See Preface for details.

Content Development: Bob Carroll
Content Management: Jeff Weidenaar, Jonathan Krebs
Content Production: Lauren Morse, Nicholas Sweeney
Product Management: Steve Schoen
Product Marketing: Alicia Wilson
Rights and Permissions: Tanvi Bhatia, Anjali Singh

Please contact https://support.pearson.com/getsupport/s/ with any queries on this content

Cover Image by Anastasia Osipova/Shutterstock.

Cover design by Jerilyn DiCarlo

Copyright © 2022, 2018, 2014 by Pearson Education, Inc. or its affiliates, 221 River Street, Hoboken, NJ 07030. All Rights Reserved. Manufactured in the United States of America. This publication is protected by copyright, and permission should be obtained from the publisher prior to any prohibited reproduction, storage in a retrieval system, or transmission in any form or by any means, electronic, mechanical, photocopying, recording, or otherwise. For information regarding permissions, request forms, and the appropriate contacts within the Pearson Education Global Rights and Permissions department, please visit www.pearsoned.com/permissions/.

Acknowledgments of third-party content (See Student Edition textbook), which constitutes an extension of this copyright page.

PEARSON, ALWAYS LEARNING, and MYLAB are exclusive trademarks owned by Pearson Education, Inc. or its affiliates in the U.S. and/or other countries.

Unless otherwise indicated herein, any third-party trademarks, logos, or icons that may appear in this work are the property of their respective owners, and any references to third-party trademarks, logos, icons, or other trade dress are for demonstrative or descriptive purposes only. Such references are not intended to imply any sponsorship, endorsement, authorization, or promotion of Pearson's products by the owners of such marks, or any relationship between the owner and Pearson Education, Inc., or its affiliates, authors, licensees, or distributors.

Library of Congress Cataloging-in-Publication Data
Cataloging-in-Publication Data is available on file at the Library of Congress.

36 2023

Activity Manual
ISBN-10: 0-13-693756-X
ISBN-13: 978-0-13-693756-2

CONTENTS

The Class Activities were written by Sybilla Beckmann as an essential part of her textbook, *Mathematics for Elementary and Middle School Teachers*. They are central and integral to full comprehension. These activities, which were previously housed in the back of the textbook itself, are with this edition available to students in this consumable workbook format (ISBN: 978-0-13-693756-2) and downloadable within MyLab Math and via the QR codes in the textbook. The fact that the Activities have, in this edition, been moved out of the Student Edition should not be taken as a sign of decreased importance—on the contrary, we have tried to make them even more accessible and useful.

All good teachers of mathematics know mathematics is not a spectator sport. We must actively think through mathematical ideas to make sense of them for ourselves. When students work on problems in the class activities—first on their own, then in a pair or a small group, and then within a whole-class discussion—they have a chance to think through the mathematical ideas several times. By discussing mathematical ideas and explaining their solution methods to each other, students can deepen and extend their thinking. As every mathematics teacher knows, students really learn mathematics when they have to explain it to someone else.

A number of activities and problems offer opportunities to critique reasoning. For example, in Class Activities 2-S, 3-O, 7-A, 7-O, 12-R, 14-T, 15-E, and 16-B, students investigate common errors in comparing fractions, adding fractions, distinguishing proportional relationships from those that are not, determining perimeter, working with similar shapes, displaying data, and probability. Since most misconceptions have a certain plausibility about them, it is important to understand what makes them mathematically incorrect. By examining what makes misconceptions incorrect, teachers deepen their understanding of key concepts and principles, and they develop their sense of valid mathematical reasoning. I also hope that, by studying and analyzing these misconceptions, teachers will be able to explain to their students why an erroneous method is wrong, instead of just saying, "You can't do it that way."

Additional Notes

- Class Activities now contain a **Materials list** at the beginning that calls attention to any necessary resources outside of the worksheet itself. Many of these resources are available as downloadable pages (**bit.ly/2SWWFUX**), ready for printing or photocopying.

- In this edition we combined the separate **indexes and bibliographies** for the textbook itself and the Class Activities into one index and one bibliography, located in the Student Edition (and Instructor's Edition). The page references for all Class Activity pages are differentiated with the prefix "CA."

- **Additional Activities** (downloadable by chapter in MyLab Math) consist of activities not included in the manual.

SECTION 1.1   CLASS ACTIVITY 1-A

The Counting Numbers as a List

CCSS   CCSS SMP3, SMP7

One way to think about the counting numbers is as a list. What are the characteristics of this list?

1. Let's first examine errors that very young children commonly make when they are first learning to say the list of counting numbers. Here are some examples of these errors:

   a. Child 1 says: "1, 2, 3, 4, 5, 8, 9, 4, 5, 2, 6, . . ."

   b. Child 2 says: "1, 2, 3, 1, 2, 3, . . ."

   Identify the nature of the errors. What are characteristics of the correct list of counting numbers?

2. Children usually learn the list of counting numbers at around the same time they learn the alphabet. Compare and contrast the alphabet and the counting numbers. In particular, why is the order of the list of counting numbers more important than the order of the letters of the alphabet?

SECTION 1.1 CLASS ACTIVITY 1-B 🍎

Connecting Counting Numbers as a List with Cardinality

CCSS CCSS SMP2, K.CC.4

To determine the number of objects in a set, we generally count the objects one by one. The process of counting the objects in a set connects the list view of the counting numbers with cardinality. Surprisingly, this connection is more subtle and intricate than we might think.

1. Spend a moment thinking about this question: If a child can correctly say the first five counting numbers—"one, two, three, four, five"—will the child necessarily be able to determine how many blocks are in a collection of 5 blocks? Why or why not? Return to this question after completing parts 2 and 3.

2. Let's examine some errors that very young children commonly make when they are first learning to count the number of objects in a set. Examples of errors follow. The picture of a pointing hand indicates a child pointing to the object. A number indicates a child saying the number.

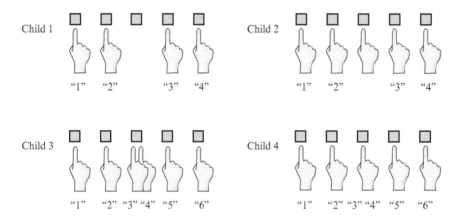

What are characteristics of correctly counting a set of objects and how does this process connect the counting numbers as a list with cardinality?

3. Compare the responses of the following two children to a teacher's request to determine how many blocks there are. Even though both children make a one-to-one cor-
respondence between the 5 blocks and the list 1, 2, 3, 4, 5, do both children appear to
understand counting equally well? If not, what is the difference?

Teacher: "How many blocks are there?"

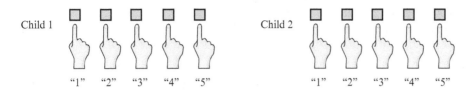

Teacher: "So how many blocks are there?"

4. Return to the question in part 1 of this activity.

5. Compare the children's responses to the teacher's question in the next scenario.

The teacher shows a child some toy bears:

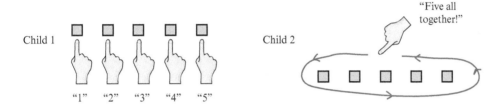

The child counts 6 bears. Then the teacher covers the bears and puts
one more bear to the side:

The teacher says: "Now how many bears are there in all?"

a. Child 1 is unable to answer.

b. Child 2 says "1, 2, 3, 4, 5, 6" while pointing to the covered bears, then points at the new
bear and says "7," and finally says, "There are 7 bears."

c. Child 3 says "6" while pointing to the covered bears, then points at the new bear and
says "7," and finally says, "There are 7 bears."

6. Discuss how Child 2 and Child 3 went *back and forth* between the "list" and "cardinality"
views of the counting numbers to answer the teacher's question in part 5.

SECTION 1.1 CLASS ACTIVITY 1-C

How Many Are There?

CCSS   CCSS SMP7, 1.NBT.2, 2.NBT.1

Materials Between 25 and 90 toothpicks (or coffee stirrers) for each person, so that the entire class has between 1000 and 2000 toothpicks. About 250 rubber bands for the class.

The purpose of this activity is to help you think about base-ten structure and our way of writing numbers.

1. Arrange your toothpicks so that you can *visually see* how many toothpicks you have. Use rubber bands to help you organize your toothpicks. Show how you arranged your toothpicks.

2. Does the way you arranged your toothpicks in part 1 correspond to the way you write the number that represents how many toothpicks you have? If so, explain how. If not, arrange them so that they do.

3. Put your toothpicks together with the toothpicks of several other people. Once again, arrange the toothpicks to correspond to the way we write the number that stands for how many toothpicks there are. Use rubber bands to help organize the toothpicks. Show how you arranged the toothpicks.

4. Repeat part 3 but now with the toothpicks from everyone in the class. How many toothpicks are there in all?

SECTION 1.1   CLASS ACTIVITY 1-D

What Do the Digits in a Counting Number Mean?

1. Suppose a child counts that there are 23 counting chips in a collection. You then write the number 23 and ask the child to show you what the 2 in 23 stands for. The child shows you 2 of the counting chips.

   What does the 2 actually stand for? How could you show that with the chips? Could you organize the chips to show the 2 as 2 of *something*?

2. **a.** Make a math drawing showing how to bundle 137 toothpicks so that the way the toothpicks are organized corresponds to the way we write the number 137.

   b. What does the 3 in 137 mean? Use your drawing from part (a) to explain.

   c. What does the 1 in 137 mean? Use your drawing from part (a) to explain.

3. **a.** In the number 2547, what does the 5 stand for?

   b. Does the 5 in 2547 stand for 5 of something? If so, 5 of what?

   c. Does the 5 in 2547 stand for 50 of something? If so, 50 of what?

4. Examine the math drawing of bagged and loose toothpicks below. Explain why the way the toothpicks are organized does not correspond to the way we write that number of toothpicks. Show how to reorganize these bagged and loose toothpicks to do so.

SECTION 1.1 **CLASS ACTIVITY 1-E**

Counting in Other Bases

CCSS CCSS SMP7, SMP8

Materials Each pair will need 40 toothpicks (or other small objects) and 18 rubber bands. If the materials are not available, drawings can be made instead.

In this activity, you will count toothpicks in base 4, base 5, and base 3, to help you better appreciate how place value works. Other bases work just like base ten except with different units. The base-ten units are ones, tens, hundreds, and so on, where each unit is 10 times the previous unit. The base-four units are ones, fours, sixteens, and so on, where each unit is 4 times the previous unit. In general, in base N, each unit is N times the previous unit.

Toothpicks **bundled** to show base-four structure:	Writing in base four:	Speaking in base four:
|	1_4	1 one
||	2_4	2 ones
|||	3_4	3 ones
⧼⧽	10_4	1 four (and 0 ones)
⧼⧽ |	11_4	1 four and 1 one
⧼⧽ ||	12_4	1 four and 2 ones
⧼⧽ |||	13_4	1 four and 3 ones
⧼⧽ ⧼⧽	20_4	2 fours (and 0 ones)
⧼⧽ ⧼⧽ |	21_4	2 fours and 1 one

1. Examine the base-four examples above. Discuss what you notice with a partner.

2. Work with a partner. One person will count your collection of toothpicks one by one in base four, saying each number in base four and bundling the toothpicks to show base-four structure. The other person will write each number of toothpicks in base four as the number is said. Follow the example on the previous page.

3. Unbundle the toothpicks. Then repeat part 2, but this time in base five.

4. Unbundle the toothpicks. Then repeat part 2, but this time in base three.

5. Unbundle the toothpicks. Then repeat part 2, but this time in another base of your choice (other than base ten).

6. What was difficult about counting in other bases? What insights do you now have about place value?

SECTION 1.2   **CLASS ACTIVITY 1-F**

Representing Decimals with Bundled Objects

CCSS   CCSS SMP2, SMP5

1. Let this large square represent *one*:

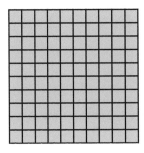

   a. What does this represent? Explain!

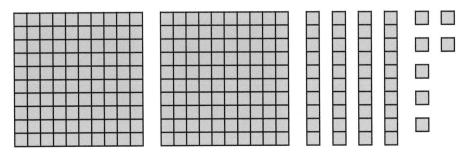

   b. Given that the large square represents *one*, make a drawing to show 3.08.

   c. Given that the large square represents *one*, make a drawing to show 0.6.

2. Let 1 paperclip represent a *thousandth*. Using this way of representing a thousandth, make simple math drawings or describe how to represent:

a. a *hundredth*,

b. a *tenth*, based on how you drew a hundredth,

c. *one*, based on how you drew a tenth,

d. 0.034,

e. 0.257,

f. 0.61.

3. Refer to your work in part 2.

  a. In the decimal 0.257, what does the 5 mean? Is there another way to interpret it?

  b. In the decimal 0.257, what does the 2 mean? Is there another way to interpret it?

4. Let's let 1 small bead stand for 0.0001, or one ten-thousandth. Show simple drawings of bundled beads so that the way of organizing the beads corresponds to the way we write the following decimals:
0.0028

0.012

5. List at least three different decimals that the toothpicks pictured below could represent. In each case, state the value of the single toothpick.

SECTION 1.2 CLASS ACTIVITY 1-G

Representing Decimals as Lengths

CCSS   CCSS SMP2, SMP5

Materials Download 1-1 at bit.ly/2SWWFUX, scissors, and tape.

A good way to represent positive decimal numbers is as lengths. Cut out the 5 long strips on Download 1-1 and tape them end-to-end without overlaps to make one long strip. The length of this long strip is 1 unit. Cut out the ten tenth strips.

1. By placing strips end-to-end without gaps or overlaps, answer the following:
    a. How many tenth strips does it take to make the 1-unit-long strip?

    b. How many hundredth strips does it take to make a tenth strip?

    c. How many thousandth strips does it take to make a hundredth strip?

    Now cut apart the hundredth and thousandth strips.

2. Represent the following decimals as lengths by placing appropriate strips end-to-end without gaps or overlaps (as best you can). In each case, draw a rough sketch (which need not be to scale) to show how you represented the decimal as a length.
    a. 1.234

    b. 0.605

    c. 1.07

    d. 1.007

3. A student was asked to place tenths on this number line:

    The student put ten tick marks between 0 and 1:

What's wrong with that?

SECTION 1.2 **CLASS ACTIVITY 1-H**

Zooming In on Number Lines

CCSS   CCSS SMP5, SMP7, SMP8

1. Label the tick marks on the number lines that follow with appropriate decimals. The second, third, and fourth number lines should be labeled as if they are "zoomed in" on the indicated portion of the previous number line.

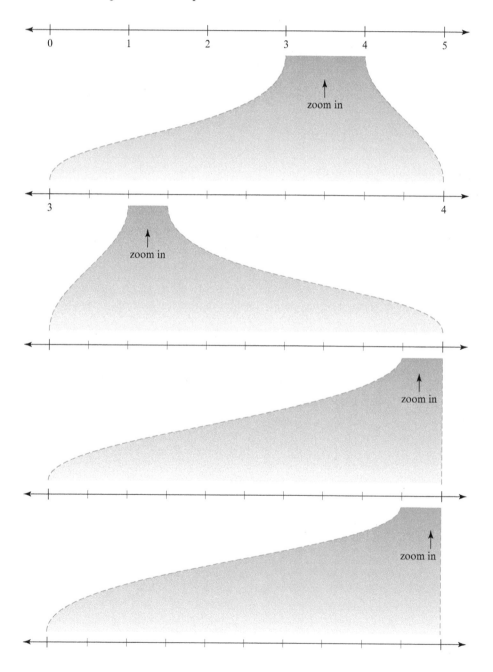

2. Now plot 3.2996 on each of the number lines in part 1 (it's easiest to start at the last number line and work backward). Use the number lines to answer the following questions:

a. Which whole numbers does 3.2996 lie between?

b. Which tenths does 3.2996 lie between?

c. Which hundredths does 3.2996 lie between?

3. Label the tick marks on the next three number lines in three different ways. In each case, the distance between adjacent tick marks should be a base-ten unit. It may help you to think about zooming in on the number line.

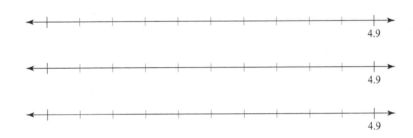

4. Label the tick marks on the next three number lines in three different ways. In each case, the distance between adjacent tick marks should be a base-ten unit. It may help you to think about zooming in on the number line.

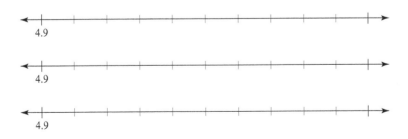

SECTION 1.2 | **CLASS ACTIVITY 1-I**

Numbers Plotted on Number Lines

CCSS CCSS SMP3, SMP7

1. What number could the point labeled A on the next number line be? Among the numbers in this list, which ones could A possibly be? Which ones could A definitely not be? Why?

$$1.14, \quad 1.3915, \quad 1.834, \quad 1.4, \quad 1.4263, \quad 1.43, \quad 1.644$$

2. Label the tick marks on the following number lines so that the tick marks are as specified, and so that the given number can be plotted on the number line. The number need not land on a tick mark.

Plot 23.84

Long ticks: whole numbers
Short ticks: tenths

Plot 0.03402

Long ticks: hundredths
Short ticks: thousandths

Plot 0.005

Long ticks: tenths
Short ticks: hundredths

Plot 7.0095

Long ticks: whole numbers
Short ticks: tenths

3. Label the tick marks on the next number lines appropriately (so that the long and short tick marks fit with the structure of the base-ten system) and explain why you labeled them that way.

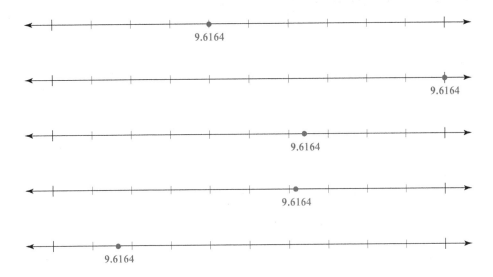

4. Hilal says that 8.571 isn't plotted correctly. She says that because the point is between the second and third tick mark to the right of a long tick mark, it should have the digit 2 in it, but 8.571 doesn't. Do you agree or disagree with Hilal? Explain!

5. Robin says that 3.2584 might be plotted correctly because it is between the fifth and sixth tick mark to the right of a long tick mark, and 3.2584 contains the digit 5. But they aren't sure yet and want to think some more. Do you agree or disagree with Robin? Explain!

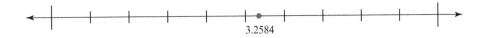

SECTION 1.2   **CLASS ACTIVITY 1-J**

Negative Numbers on Number Lines

CCSS   CCSS SMP3, 6.NS.6

1. Katie, Matt, and Parna were asked to label the tick marks on a number line on which one tick mark was already labeled as −7. What's wrong with their work? Show how to label the number line appropriately. Is there more than one way to label it appropriately?

Katie's work:

Matt's work:

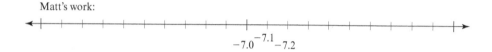

Parna's work:

Appropriate labeling:

Another appropriate labeling:

2. Sometimes students get confused about the relative locations of certain decimals, negative numbers, and zero. Plot −1 on the next number line. Then give a few examples of numbers that lie between 0 and 1 and some numbers that lie between 0 and −1. Include examples of numbers that land *between* tick marks.

3. Use a number line to explain why $-(-3) = 3$. More generally, explain why $-(-N) = N$ is true for all numbers N.

SECTION 1.3 CLASS ACTIVITY 1-K

Comparing Base-Ten Amounts

1. For each of the following pairs of amounts, determine which is greater and explain your reasoning.
   a. 4 hundreds and 18 tens and 3 ones
      5 hundreds and 7 tens and 9 ones

   b. 1 thousand
      7 hundreds and 43 tens

   c. 2 ones and 4 tenths
      1 one and 15 tenths

   d. 6 tenths
      42 hundredths

   e. 6 tenths
      83 hundredths

2. Why can we determine that 723 is greater than 699 just by comparing the 7 and the 6 in the hundreds places even though in part 1(a) it wouldn't be correct to just compare the 4 hundreds with the 5 hundreds?

3. Tenths are greater than hundredths and 1 tenth is greater than 9 hundredths, but is 1 tenth always greater than any number of hundredths? Explain!

SECTION 1.3   CLASS ACTIVITY 1-L

Critique Reasoning about Comparing Decimals

CCSS   CCSS SMP3

The list that follows describes some common errors students make when comparing decimals. For each error, think about why students might make that error.

Error 1: 2.352 > 2.4

Error 2: 2.34 > 2.5 (but identify correctly that 2.5 > 2.06)

Error 3: 5.47 > 5.632

Error 4: 1.8 = 1.08

The next list describes some of the misconceptions students can develop about comparing decimals. (See [Sta05] and [Ste02] for further information, including additional misconceptions and advice on instruction.)

Whole number thinking: Students with this misconception treat the portion of the number to the right of the decimal point as a whole number, thus thinking that 2.352 > 2.4 because 352 > 4. These students think that longer decimals are always larger than shorter ones.

Column overflow thinking: Students with this misconception name decimals incorrectly by focusing on the first nonzero digit to the right of the decimal point. For example, they say that 2.34 is "two and thirty-four tenths." These students think that 2.34 > 2.5 because 34 tenths is more than 5 tenths. These students usually identify longer decimals as larger; they will, however, correctly identify 2.5 as greater than 2.06 because 5 tenths is more than 6 hundredths.

Denominator-focused thinking: Students with this misconception think that any number of tenths is greater than any number of hundredths and that any number of hundredths is greater than any number of thousandths, and so on. These students identify 5.47 as greater than 5.632, reasoning that 47 hundredths is greater than 632 thousandths because hundredths are greater than thousandths. Students with this misconception identify shorter decimal numbers as larger.

Reciprocal thinking: Students with this misconception view the portion of a decimal to the right of the decimal point as something like the fraction formed by taking the reciprocal. For example, they view 0.3 as something like $\frac{1}{3}$ and thus identify 2.3 as greater than 2.4 because $\frac{1}{3} > \frac{1}{4}$. These students usually identify shorter decimal numbers as larger, except in cases of intervening zeros. For example, they may say that 0.03 > 0.4 because $\frac{1}{3} > \frac{1}{4}$.

Money thinking: Students with this difficulty truncate decimals after the hundredths place and view decimals in terms of money. If two decimals agree to the hundredths place, these students simply guess which one is greater—sometimes guessing correctly, sometimes guessing incorrectly. Most of these students recognize that 1.8 is like $1.80, although some view 1.8 incorrectly as $1.08.

SECTION 1.3 CLASS ACTIVITY 1-M

Finding Decimals between Decimals

CCSS CCSS SMP8

1. Contemplate the questions in the following paragraph for a few minutes before you continue with the rest of the activity. Then return to these questions at the end.

   There aren't any whole numbers between 2 and 3, but there are plenty of decimals between 2 and 3. If you are given two decimals, will there always be another decimal between the two? Are there some decimals that don't have any other decimals between them?

2. Work with a partner and take turns giving your partner a pair of decimals and challenging them to find a decimal between your pair. For example, you could give your partner the pair 1.2, 1.4; your partner could respond with 1.3. Try to stump your partner! Continue taking turns until one of you has stumped the other or you both agree that neither of you will be able to stump the other.

3. Return to the questions in part 1. Have your answers changed? How?

Decimals between Decimals on Number Lines

For each of the pairs of numbers that follow, find a number between the two numbers. Label the longer tick marks on the number line so that all three numbers can be plotted visibly and distinctly. The labeling should fit with the structure of the base-ten system. Plot all three numbers. The numbers need not land on tick marks.

1. The numbers 1.6 and 1.7

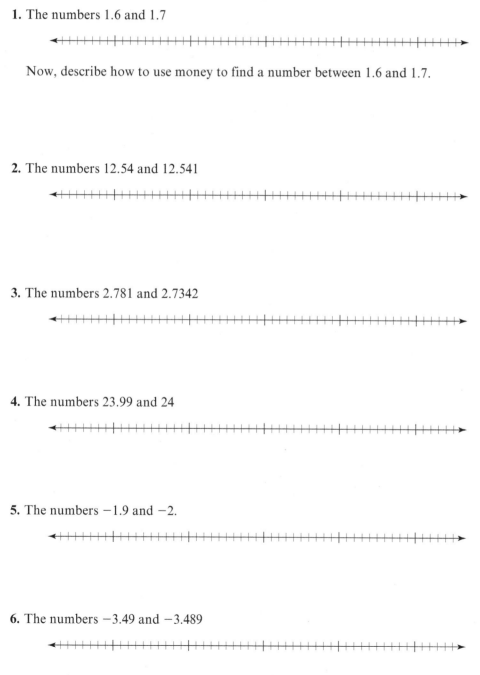

   Now, describe how to use money to find a number between 1.6 and 1.7.

2. The numbers 12.54 and 12.541

3. The numbers 2.781 and 2.7342

4. The numbers 23.99 and 24

5. The numbers −1.9 and −2.

6. The numbers −3.49 and −3.489

SECTION 1.3 CLASS ACTIVITY 1-O

"Greater Than" and "Less Than" with Negative Numbers

CCSS CCSS SMP2, SMP7

1. Explain in two different ways why a negative number is always less than a positive number.

2. A student might think that $-5 > -2$. Describe two different ways to explain why this is not correct.

3. For each of the following pairs of numbers, write an inequality to show how the numbers compare. Explain your reasoning.
   a. -6 and -5.9

   b. -4.777 and -4.8

   c. -0.555 and -0.1

4. Given the following number line, write inequalities to show how each of the pairs of numbers below compare. Explain your reasoning.

   a. $-R$ and $-P$

   b. R and P

SECTION 1.4   **CLASS ACTIVITY 1-P**

Explaining Rounding

CCSS   CCSS 3.NBT.1

1. Round points *A* and *B* to the nearest hundred.

What might the base-ten representations of *A* and *B* be? Why?

2. Which numbers in the interval between 7300 and 7400 round to 7300, when rounding to the nearest hundred? Indicate that range of numbers on the number line.

What characterizes the base-ten representations of those numbers?

3. Which numbers in the interval between 7300 and 7400 round to 7400, when rounding to the nearest hundred? Indicate that range of numbers on the number line.

What characterizes the base-ten representations of those numbers?

4. When we round a number to the nearest hundred, why do we look to the tens place? What does the digit in the tens place tell us about the location of the number on a number line labeled with tick marks representing hundreds?

SECTION 1.4   CLASS ACTIVITY 1-Q

Rounding with Number Lines

CCSS CCSS 5.NBT.4

Using number lines to round can help us focus on the base-ten system and understand its structure better.

1. The tick marks on the following number line are labeled with thousands. Plot 38721 in its approximate location on this number line. Then use the number line to explain how to round 38721 to the nearest thousand.

2. Label the unlabeled tick marks on the following number line with appropriate tens so that you can plot 5643 in its approximate location on this number line. Then use the number line to explain how to round 5643 to the nearest ten.

3. The tick marks on the following number line are labeled with hundredths. Plot 2.349 in its approximate location on this number line. Then use the number line to explain how to round 2.349 to the nearest hundredth.

4. Label the unlabeled tick marks on the following number line with appropriate tenths so that you can plot 2.349 in its approximate location on this number line. Then use the number line to explain how to round 2.349 to the nearest tenth.

5. Label the tick marks on the following number line so that you can use the number line to explain how to round 54,831 to the nearest hundred.

6. Label the tick marks on the following number line so that you can use the number line to explain how to round 16.936 to the nearest hundredth.

7. Label the tick marks on the following number line so that you can use the number line to explain how to round 16.936 to the nearest tenth.

SECTION 1.4   CLASS ACTIVITY 1-R

Can We Round This Way?

CCSS CCSS SMP3

Maureen has made up her own method of rounding. Starting at the rightmost place in a decimal number, she keeps rounding to the value of the next place to the left until she reaches the place to which the decimal number was to be rounded.

For example, Maureen would use the following steps to round 3.2716 to the nearest tenth:

$$3.2716 \rightarrow 3.272 \rightarrow 3.27 \rightarrow 3.3$$

Try Maureen's method on several examples. Is her method valid? That is, does it always round decimal numbers correctly? Or are there examples of decimal numbers where Maureen's method does not give the correct rounding?

How Can We Define Fractions?

1. Off the top of your head, what is one way to describe what a fraction such as $\frac{3}{5}$ means?

2. Can you think of other ways to interpret what $\frac{3}{5}$ means?

3. One common way to interpret a fraction $\frac{A}{B}$ is "*A* out of *B* equal parts." This is correct and appropriate in some situations, but it can cause difficulties in other situations. For each of the following, discuss how interpreting a fraction $\frac{A}{B}$ as "*A* out of *B* equal parts" could cause errors or lead to difficulties.

a. Interpret the fraction $\frac{7}{5}$.

b. Plot $\frac{3}{5}$ on the number line.

c. Add $\frac{4}{5}$ and $\frac{3}{5}$.

SECTION 2.1   CLASS ACTIVITY 2-B

Getting Familiar with Our Definition of Fractions

CCSS   CCSS SMP6, 3.NF.1

1. Work with a partner. Use *our definition of fraction given in the text* to describe the three fractions on the left (or the right) to your partner. Listen to your partner's description of the other three fractions.

Unit amount or whole

Unit amount or whole

Show $\frac{1}{3}$ of the unit amount:

Show $\frac{2}{3}$ of the unit amount:

Show $\frac{4}{3}$ of the unit amount:

Show $\frac{1}{5}$ of the unit amount:

Show $\frac{3}{5}$ of the unit amount:

Show $\frac{6}{5}$ of the unit amount:

2. For each of the following, use our definition of fraction to interpret the given strip. Then draw the requested unit amount and explain your answer.

   a. $\frac{3}{4}$ cup butter:

     Interpret the strip as $\frac{3}{4}$:
     Draw 1 cup butter:

   b. $\frac{3}{5}$ of a liter of juice:

     Interpret the strip as $\frac{3}{5}$:
     Draw 1 liter of juice:

   c. $\frac{4}{9}$ of a pound of flour:

     Interpret the strip as $\frac{4}{9}$:
     Draw 1 pound of flour:

   d. $\frac{6}{5}$ of a kilogram of sugar:

     Interpret the strip as $\frac{6}{5}$:
     Draw 1 kilogram of sugar:

3. Discuss how a 3 in the numerator of a fraction $\frac{3}{\bigcirc}$ is like a digit 3 in a number 3◌◌ in base ten. Also, how are unit fractions like base-ten units?

SECTION 2.1 **CLASS ACTIVITY 2-C** 🍎

Using Our Fraction Definition to Solve Problems

CCSS CCSS SMP1, SMP6

Materials You will need pattern tiles for parts 5 and 6. (You can cut out pattern tiles from Downloads G-1 and G-2 at bit.ly/2SWWFUX.)

Use our definition of fractions throughout and notice that it can help you solve these problems.

1. Take a blank piece of paper and imagine that it is $\frac{4}{5}$ of some larger piece of paper. Fold your piece of paper to show $\frac{3}{5}$ of the larger (imagined) piece of paper. Do this as carefully and precisely as possible without using a ruler or doing any measuring. Explain why your answer is correct. Could two people have different-looking solutions that are both correct?

2. Benton used $\frac{3}{4}$ cup of butter to make a batch of cookie dough. He rolled his cookie dough out into a rectangle. Now Benton wants a portion of the dough that contains $\frac{1}{4}$ cup of butter. How could Benton cut the dough? Explain your answer.

3. The strip below shows a greenway that is $\frac{9}{8}$ of a mile long. Partition the strip and use it as a guide to draw another greenway that is 1 mile long. Explain your reasoning.

4. The strip below shows a path that is $\frac{3}{4}$ of a kilometer long. Use the strip as a guide to draw another path that is 1 kilometer long. Explain your reasoning.

5. a. The yellow hexagon pattern tile is $\frac{2}{3}$ of the area of a pattern tile design. Use pattern tiles to make the design (what it could be). Explain your reasoning.

   b. What if the yellow hexagon is $\frac{3}{2}$ of the area of another design?

   c. What if the yellow hexagon is $\frac{6}{10}$ of the area of another design?

6. Suppose 3 yellow hexagons are $\frac{2}{7}$ of the area of a pattern tile design. Use pattern tiles to make the design (what it could be). Explain your reasoning.

SECTION 2.1 CLASS ACTIVITY 2-D

Why Are Fractions Numbers? A Measurement Perspective

CCSS CCSS SMP2, SMP8

1. Contemplate and discuss the following question: Why are fractions numbers? For example, why are $\frac{2}{3}$ and $\frac{8}{3}$ numbers just like 2 and 3 are numbers?

2. How many of the unit strip does it take to make Strip X exactly?
   How many of the unit strip does it take to make Strip Y exactly?
   How many of the unit strip does it take to make Strip Z exactly?
   How many of the unit strip does it take to make Strip W exactly?

3. Return to part 1 and discuss it some more.

We can think of numbers as the result of measuring by a unit amount or a 1. If you have 4 pounds of flour, the 4 tells you how many "1 pounds" you have. Similarly, if you have $\frac{1}{4}$ of a pound of flour, the $\frac{1}{4}$ tells you how many (i.e., how much) of "1 pound" you have. Use this measurement perspective *and our definition of fraction* as you relate quantities in the following problems.

4. a. How many of Strip A does it take to make Strip B?

    b. How many of Strip B does it take to make Strip A? Explain briefly.

5. a. How many of Strip C does it take to make Strip D? Explain.

    b. How many of Strip D does it take to make Strip C? Explain.

6. a. How many of Strip E does it take to make Strip F? Explain.

    b. How many of Strip F does it take to make Strip E? Explain.

7. For each of 4–6, what do you notice about the fractions in parts (a) and (b)?

SECTION 2.1 **CLASS ACTIVITY 2-E**

Relating Fractions to Wholes

CCSS CCSS SMP1, SMP6

1. At a neighborhood park, $\frac{1}{3}$ of the area of the park is to be used for a new playground. Swings will be placed on $\frac{1}{4}$ of the area of the playground. What fraction of the neighborhood park will the swing area be?

   a. Make a math drawing to help you solve the problem and explain your solution. Use our definition of fraction in your explanation.

   b. For each fraction that appears in the problem and its solution, determine its unit amount (whole). Ask yourself, "This fraction is a fraction of *what*?"

   c. Discuss how one amount can be described with two different fractions depending on what the unit amount is taken to be.

2. Ben is making a recipe that calls for $\frac{1}{3}$ cup of oil. Ben has a bottle that contains $\frac{2}{3}$ cup of oil. Ben does not have any measuring cups. What fraction of the oil in the bottle should Ben use for his recipe?

   a. Make a math drawing to help you solve the problem and explain your solution. Use our definition of fraction in your explanation.

   b. For each fraction that appears in the problem and its solution, determine its unit amount (whole). Ask yourself, "This fraction is a fraction of *what*?"

   c. Discuss how one amount can be described with two different fractions depending on what the unit amount is taken to be.

SECTION 2.1   CLASS ACTIVITY 2-F

Critiquing Fraction Arguments

CCSS CCSS SMP3, SMP6

The unit amount that a fraction refers to need not be a single contiguous object. Instead, the unit amount can consist of several pieces that need not even be the same size. Working with a noncontiguous unit amount provides an opportunity to think more deeply about the definition of fraction.

Recall these definitions:

- An amount is $\frac{1}{B}$ of a unit amount if B copies of it joined together are the same size as the unit amount.

- An amount is $\frac{A}{B}$ of a unit amount if it can be formed by A parts, each of which is $\frac{1}{B}$ of the unit amount.

1. *Peter's garden problem:* The drawing below is a map of Peter's garden, which consists of two plots. The two plots have each been partitioned into 5 pieces of equal area. The shaded parts show where carrots have been planted. What fraction of the area of Peter's garden is planted with carrots?

   Spend a few minutes solving this problem yourself. Then move on to the next parts, which show methods that some students used as they attempted to solve this problem.

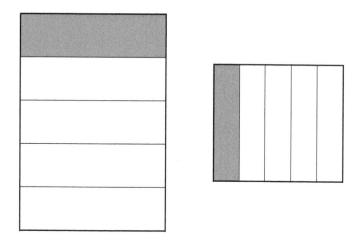

2. Mariah thought about Peter's garden problem this way: "There are 2 parts out of a total of 10 parts, so $\frac{2}{10}$ of the garden is planted with carrots. Since $\frac{2}{10} = \frac{1}{5}$, then $\frac{1}{5}$ of the garden is planted with carrots."

   Is Mariah's reasoning valid? Why or why not?

   If the two garden plots had been the same size, then would her reasoning be valid?

3. Aysah drew this picture as she was thinking about Peter's garden problem:

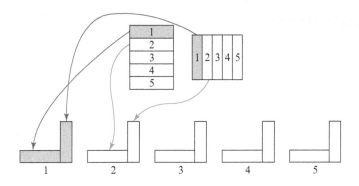

Can Aysah's picture be used to solve the problem? Explain.

4. According to Matt, $\frac{1}{5}$ of the first plot is shaded and $\frac{1}{5}$ of the second plot is shaded. Because there are two parts shaded, each of which is $\frac{1}{5}$, this means $\frac{2}{5}$ of Peter's garden is planted with carrots.

Is Matt's reasoning valid? Why or why not?

What if the two garden plots were the same size and each plot had an area of 1 acre? Could Matt's reasoning be used to make a correct statement? Explain.

SECTION 2.1   CLASS ACTIVITY 2-G

Critique Fraction Locations on Number Lines

CCSS   CCSS SMP6, 3.NF.2

1. Students were asked to plot the fractions $\frac{1}{4}, \frac{2}{4}$, and $\frac{3}{4}$ on a number line.

   Eric's work:

   Kristin's work:

   How might Eric be thinking? Although Kristin's labeling is not incorrect, what might she not be attending to?

2. When Tyler was asked to plot $\frac{3}{4}$ on a number line showing 0 and 2, he plotted it as shown on the next number line. How might Tyler be thinking?

   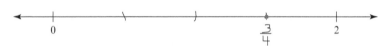

3. When Amy was asked what fraction should go in the box, she wrote $\frac{2}{6}$. Why might she have done so? What idea might she not be attending to?

4. Discuss ideas you have for helping students understand number lines. How might you draw their attention to the lengths of intervals and to distance from 0 and away from merely counting tick marks without attending to length and distance?

SECTION 2.1 **CLASS ACTIVITY 2-H** 🍎

Fractions on Number Lines

CCSS CCSS SMP1, SMP3, 3.NF.2

1. Explain in detail how to determine where to plot $\frac{3}{4}$ and $\frac{5}{4}$ on the number line below, and explain why those locations fit with our definition of fraction.

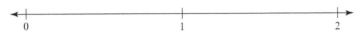

2. For each of the following problems, place equally spaced tick marks on the number line so that you can plot the requested fraction on a tick mark. You may place the tick marks "by eye"; precision is not needed. Explain your reasoning.
    a. Plot $\frac{5}{4}$.

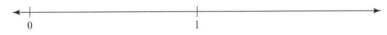

    b. Plot 1.

    c. Plot 1.

    d. Plot $\frac{3}{5}$.

SECTION 2.1   CLASS ACTIVITY 2-I

Fraction Problem Solving with Pattern Tiles

CCSS   CCSS SMP1, SMP3

Materials You will need pattern tiles (yellow hexagons, red trapezoids, blue rhombuses, and green triangles only). (You can cut out pattern tiles from Downloads G-1 and G-2 at bit.ly/2SWWFUX.)

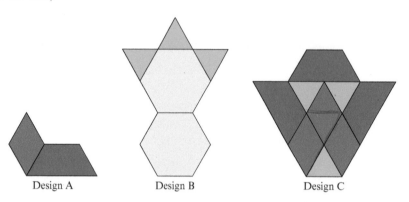

Design A   Design B   Design C

1. The area of Design A is $\frac{5}{4}$ of the area of another design. Make a design that could be the other design. Explain your reasoning.

2. The area of Design B is $\frac{5}{3}$ of the area of another design. Make a design that could be the other design. Explain your reasoning.

3. The area of Design C is $\frac{9}{7}$ of the area of another design. Make a design that could be the other design. Explain your reasoning.

Explaining Equivalent Fractions

CCSS   CCSS SMP2, SMP7, 4.NF.1

1. Why it is that when we multiply the numerator and denominator of a fraction by the same number, the resulting fraction is equal to the original one? For example, why is $\frac{2}{3}$ equal to $\frac{2 \times 4}{3 \times 4}$? Discuss some of your initial ideas with a partner.

2. Use our definition of fraction and the math drawing below to give a detailed conceptual explanation for why

$$\frac{2}{3} \text{ is equal to } \frac{2 \times 4}{3 \times 4}$$

Discuss how to use the structure of the math drawing to explain the process of multiplying both the numerator and denominator of $\frac{2}{3}$ by 4. During the process, what changes and what stays the same? Why are the two fractions *equal*?

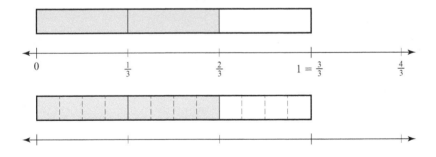

3. When you discuss equivalent fractions with students, they may find it confusing that in the math drawing, the fraction pieces are *partitioned (divided)*, but in the numerical work, we *multiply*. Address this confusing point by discussing what happens to the *size* of the pieces and what happens to the *number* of pieces when we create equivalent fractions.

4. Use a carefully structured math drawing to help you explain why $\frac{4}{5} = \frac{4 \times 3}{5 \times 3}$. In your explanation, take special care to discuss:

- what happens to the number of parts and the size of the parts;
- *how your math drawing* shows that the numerator and denominator are each multiplied by 3;
- *how your math drawing* shows why the two fractions are equal.

CLASS ACTIVITY 2-K

Critique Fraction Equivalence Reasoning

CCSS CCSS SMP3

1. Anna says that

$$\frac{2}{3} = \frac{6}{7}$$

because, starting with $\frac{2}{3}$, you get $\frac{6}{7}$ by adding 4 to the top and the bottom. If you do the same thing to the top and the bottom, the fractions must be equal.

Is Anna right? If not, why not? What should we be careful about when talking about equivalent fractions?

2. Don says that $\frac{11}{12} = \frac{16}{17}$ because both fractions are one part away from a whole. Is Don correct? If not, what is wrong with Don's reasoning?

3. Peter says that $\frac{6}{6}$ is greater than $\frac{5}{5}$ because $\frac{6}{6}$ has more parts. Is Peter correct? If not, what is wrong with Peter's reasoning?

SECTION 2.2   CLASS ACTIVITY 2-L

Interpreting and Using Common Denominators

CCSS CCSS SMP2

1. Write $\frac{3}{4}$ and $\frac{5}{6}$ with two different common denominators. In terms of the strips and the number lines, what are we doing when we give the two fractions common denominators? During the process, what changes and what stays the same?

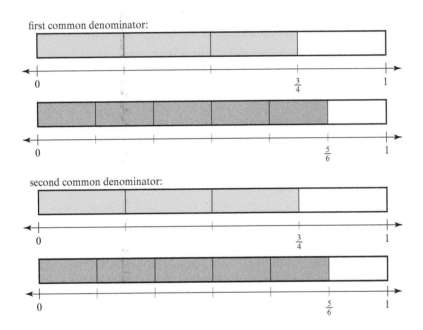

2. Plot 1, $\frac{2}{3}$, and $\frac{5}{2}$ on the number line for this problem in such a way that each number falls on a tick mark. Lengthen the tick marks of whole numbers.

3. Plot 9, $\frac{55}{6}$, and $\frac{33}{4}$ on the number line for this problem in such a way that each number falls on a tick mark. Lengthen the tick marks of whole numbers.

4. Plot 1, 0.7, and $\frac{3}{4}$ on the number line for this problem in such a way that each number falls on a tick mark. Lengthen the tick marks of whole numbers.

SECTION 2.2   CLASS ACTIVITY 2-M

Solving Problems by Using Equivalent Fractions

CCSS CCSS SMP1, SMP2

1. Take a blank piece of paper and imagine that it is $\frac{2}{3}$ of some larger piece of paper. Fold your piece of paper to show $\frac{1}{6}$ of the larger (imagined) piece of paper. Do this as carefully and precisely as possible without using a ruler or doing any measuring. Explain why your answer is correct.

   In solving this problem, how does $\frac{2}{3}$ appear as an equivalent fraction? Could two people have different solutions that are both correct?

2. Jean has a casserole recipe that calls for $\frac{1}{2}$ cup of butter. Jean only has $\frac{1}{3}$ cup of butter.

   Assuming that Jean has enough of the other ingredients, what fraction of the casserole recipe can Jean make? Make math drawings to help you solve this problem. Explain why your answer is correct. In your explanation, attend carefully to the unit amount that each fraction is *of*.

   In solving this problem, how do $\frac{1}{2}$ and $\frac{1}{3}$ each appear as equivalent fractions?

3. One serving of SugarBombs cereal is $\frac{3}{4}$ cup. Joey wants to eat $\frac{1}{2}$ of a serving of SugarBombs cereal. How much of a cup of cereal should Joey eat? Make math drawings to help you solve this problem. Explain why your answer is correct. In your explanation, attend carefully to the unit amount that each fraction is *of*.

   In solving the problem, how does $\frac{3}{4}$ appear as an equivalent fraction?

SECTION 2.2 **CLASS ACTIVITY 2-N**

Problem Solving with Common Partitioning

Be sure to draw your partitioning lines so that you can see the difference between the original parts of each strip and the new mini-parts that you create within each part.

1. Partition the strips below into mini-parts so that all the mini-parts are identical.

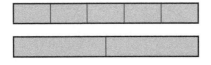

How many mini-parts did you make in each of the 5 parts of the top strip?

How many mini-parts did you make in each of the 2 parts of the bottom strip?

How are the numbers of parts and mini-parts are related in each of the two strips?

2. Partition the strips below into mini-parts so that all the mini-parts are identical. Anticipate how to partition before you do so.

How many mini-parts should you make in each of the 5 parts of the top strip?

How many mini-parts should you make in each of the 3 parts of the bottom strip?

Why will it make sense to do the partitioning that way?

3. Partition the strips below into mini-parts so that all the mini-parts are identical. Anticipate how to partition before you do so.

How many mini-parts should you make in each of the 4 parts of the top strip?

How many mini-parts should you make in each of the 6 parts of the bottom strip?

Why will it make sense to do the partitioning that way?

Is there another way you could do the partitioning?

4. Your workout was to row $\frac{2}{3}$ of a mile, but you rowed only $\frac{1}{2}$ of a mile. What fraction of your workout did you do?

See if you can solve this workout problem by reasoning about how to partition the strips below!

5. Partition the strips below into mini-parts so that all the mini-parts are identical. Notice that 5 parts of the top strip are aligned with 2 parts of the bottom strip.

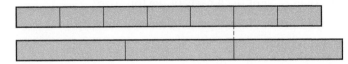

How many mini-parts did you make in each of the parts of the top strip?

How many mini-parts did you make in each of the parts of the bottom strip?

Why did it make sense to partition that way?

6. Partition the strips below into mini-parts so that all the mini-parts are identical. Notice that 5 parts of the top strip are aligned with 3 parts of the bottom strip. Anticipate how to partition before you do so.

How many mini-parts should you make in each of the parts of the top strip?

How many mini-parts should you make in each of the parts of the bottom strip?

Why will it make sense to do the partitioning that way?

7. A wire is $\frac{2}{3}$ of a meter long. How long is $\frac{1}{3}$ of the wire?

See if you can solve this wire problem by reasoning about how to partition the strips below!

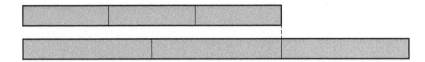

8. Swimming at a steady pace, it took Emerson $\frac{3}{4}$ of an hour to swim $\frac{5}{8}$ of a mile. If Emerson swims at that same pace, how far will they swim in 1 hour?

See if you can solve this swimming problem by reasoning about how to partition the strips below!

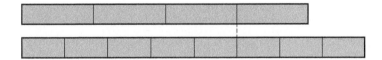

SECTION 2.2 **CLASS ACTIVITY 2-O**

Problem Solving with Fractions on Number Lines

CCSS CCSS SMP1, SMP2

Explain how to solve the following problems by *reasoning about partitioning and equivalent fractions*. Place equally spaced tick marks on the number line so that the given fraction and the requested fraction both land on tick marks. You may place the tick marks "by eye"; precision is not needed.

1. Plot $\frac{3}{4}$ without first plotting 1.

2. Plot $\frac{3}{5}$ without first plotting 1.

3. Plot $\frac{3}{8}$ without first plotting 1.

4. So far, Sue has run $\frac{1}{4}$ of a mile, but that is only $\frac{2}{3}$ of the total distance she will run. What is her total running distance?

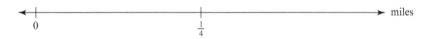

5. So far, Tyler has run $\frac{2}{3}$ of a mile, but that is only $\frac{3}{4}$ of his total running distance. What is his total running distance?

SECTION 2.2 CLASS ACTIVITY 2-P

Measuring One Quantity with Another

CCSS CCSS SMP2, SMP8

Suppose that a company makes buttons and puts them on cardboard strips.

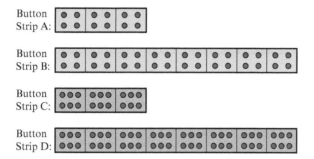

1. Discuss: What are ways to compare or relate Strips A and B? How is the relationship between Strip A and Strip B the same as the relationship between Strip C and Strip D? How is the relationship different?

2. One way to relate two quantities is to choose one quantity as a unit amount and measure the other quantity with this unit amount. Use this measurement perspective *and our definition of fraction* in the following questions.

   a. How many of Strip A does it take to have the same number of buttons as Strip B?

   b. How many of Strip B does it take to have the same number of buttons as Strip A?

   c. How many of Strip C does it take to have the same number of buttons as Strip D?

   d. How many of Strip D does it take to have the same number of buttons as Strip C?

   e. If you didn't already, explain how to answer part (a) with two equivalent fractions. Likewise, explain how to answer each of parts (b) through (d) with two equivalent fractions.

3. Return to part 1 of this activity and discuss some more!

SECTION 2.3   CLASS ACTIVITY 2-Q

What Is Another Way to Compare These Fractions?

CCSS   CCSS SMP2, SMP3, 3.NF.3d

For each of the pairs of fractions shown, determine which fraction is greater in a way other than finding common denominators, cross-multiplying, or converting to decimals. Explain your reasoning.

$$\frac{1}{49} \quad \frac{1}{39}$$

$$\frac{7}{37} \quad \frac{7}{45}$$

$$\frac{3}{23} \quad \frac{6}{49}$$

$$\frac{4}{29} \quad \frac{2}{15}$$

$$\frac{5}{201} \quad \frac{3}{110}$$

CLASS ACTIVITY 2-R

Comparing Fractions by Reasoning

CCSS   CCSS SMP2, SMP3, 4.NF.2

Use reasoning other than finding common denominators, cross-multiplying, or converting to decimals to compare the sizes ($=$, $<$, or $>$) of the following pairs of fractions:

$\dfrac{27}{43}$ $\dfrac{26}{45}$

$\dfrac{13}{25}$ $\dfrac{34}{70}$

$\dfrac{17}{18}$ $\dfrac{19}{20}$

$\dfrac{9}{40}$ $\dfrac{12}{44}$

$\dfrac{51}{53}$ $\dfrac{65}{67}$

$\dfrac{37}{35}$ $\dfrac{27}{25}$

$\dfrac{13}{25}$ $\dfrac{5}{8}$

$\dfrac{42}{57}$ $\dfrac{39}{61}$

SECTION 2.3   CLASS ACTIVITY 2-S 🍎

Can We Reason This Way?

CCSS   CCSS SMP3, 3.NF.3d, 4.NF.2

Claire says that

$$\frac{4}{9} > \frac{3}{8}$$

because

$$4 > 3 \text{ and } 9 > 8$$

Discuss whether Claire's reasoning is correct.

SECTION 2.4 **CLASS ACTIVITY 2-T**

Math Drawings, Percentages, and Fractions

1. For diagrams 1 through 5, determine the percent of each diagram that is shaded, explaining your reasoning. Write each percent as a fraction in simplest form, and explain how to see that this fraction of the diagram is shaded. You may assume that portions of each diagram that appear to be the same size really are the same size.

1

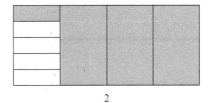

2

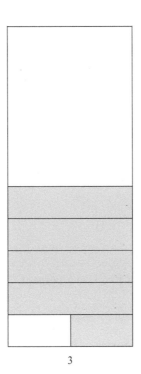

3

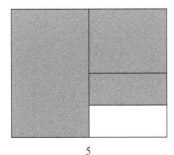

5

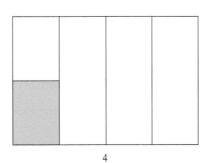

4

SECTION 2.4 **CLASS ACTIVITY 2-U** 🝙

Reasoning about Percent Tables to Solve "Portion Unknown" Percent Problems

CCSS CCSS SMP2, SMP5, 6.RP.3c

We can make some percent problems easy to solve mentally by working with benchmark percentages. Use math drawings and percent tables to help you reason and record your thinking.

1. Calculate 95% of 80,000 by calculating $\frac{1}{10}$ of 80,000 and then calculating half of that result. Use the following math drawing and the percent table to help you explain why your method makes sense and to record your thinking. Is there more than one way to find 95%?

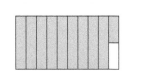

Percent table

100%	⟶	80,000
10%	⟶	_____
5%	⟶	_____
95%	⟶	_____

2. Calculate 15% of 6500 by first calculating $\frac{1}{10}$ of 6500. Use the math drawing and a percent table to help you explain why your method makes sense and to record your thinking.

Percent table

3. Calculate 7% tax on a $25 purchase by first finding 1%. Use a percent table to record the reasoning.

Percent table

100%	⟶	$25
1%	⟶	_____
7%	⟶	_____

4. Calculate 60% of 810 by first calculating $\frac{1}{2}$ of 810. Use the math drawing and a percent table to help you explain why this method makes sense.

Percent table

5. Find another way to use a percent table to calculate 60% of 810.

6. Use a percent table to calculate 55% of 180.

7. Make up a percent calculation problem, and explain how to solve it with a percent table.

SECTION 2.4   **CLASS ACTIVITY 2-V**

Reasoning about Percent Tables

CCSS   CCSS SMP1, SMP2, 6.RP.3c

1. Lenny has received 6 boxes of paper, which is 30% of the paper he ordered. How many boxes of paper did Lenny order? Make a math drawing and use a percent table to help you solve this problem. Explain your reasoning in each case.

   Math drawing: Percent table:

   $$30\% \rightarrow 6$$

2. Ms. Jones paid $2.10 in tax on an item she purchased. The tax was 7% of the price of the item. What was the price of the item (not including the tax)? Solve this problem with the aid of a percent table. Explain your reasoning.

3. In Green Valley, the average daily rainfall is typically $\frac{5}{8}$ of an inch. Last year, the average daily rainfall in Green Valley was only $\frac{3}{8}$ of an inch. What percent of the typical average daily rainfall fell last year in Green Valley? Solve this problem with the aid of either a math drawing or a percent table, or both. Explain your reasoning.

4. If a $\frac{1}{3}$ cup serving of cheese provides your full daily value of calcium, then what percentage of your daily value of calcium is provided by $\frac{3}{4}$ cup of the cheese? Solve this problem with the aid of either a math drawing or a percent table, or both. Explain your reasoning.

SECTION 2.4   **CLASS ACTIVITY 2-W**

Percent Problem Solving

CCSS CCSS SMP1

1. A bag contains 30 blue marbles, which is 40% of the marbles in the bag. How many marbles are in the bag? Solve this problem in at least one of three ways: (1) with the aid of a math drawing, (2) with the aid of a percent table, and (3) by making equivalent fractions (without cross-multiplying). Explain your reasoning.

2. Andrew ran 40% as far as Marcie. How far did Marcie run as a percentage of Andrew's running distance? Explain your answer.

3. A terrarium has contains 10% more female bugs than male bugs. If there are 8 more female bugs than male bugs, then how many bugs are in the terrarium?

   a. Solve the bug problem and explain your solution.

   b. An easy error to make is to say that there are 80 bugs in all in the terrarium. Why is this not correct and why is it an easy error to make?

4. An animal shelter has only dogs and cats. There are 25% more dogs than cats. What percentage of the animals at the animal shelter are cats? Explain your solution.

SECTION 3.1   CLASS ACTIVITY 3-A

Relating Addition and Subtraction—The Shopkeeper's Method of Making Change

CCSS   CCSS SMP7

When a patron of a store gives a shopkeeper A for a B purchase, we can think of the change owed to a patron as what is left from A when B are taken away. In contrast, the shopkeeper might make change by starting with B and handing the customer money while adding on the amounts until they reach A.

Explain how the shopkeeper's method links subtraction to addition.

SECTION 3.1   **CLASS ACTIVITY 3-B**

Writing Add To and Take From Problems

CCSS CCSS 1.OA.1, 2.OA.1

1. For each of the following equations, write a problem that is formulated naturally by the equation.

   a. $6 + 9 = ?$

   b. $6 + ? = 15$

   c. $? + 6 = 15$

   d. $15 - 6 = ?$

   e. $15 - ? = 6$

   f. $? - 6 = 9$

2. What are similarities and differences among your problems in part 1? Which problems could be solved the same way?

Writing Put Together/Take Apart and Compare Problems

CCSS   CCSS 1.OA1, 2.OA.1, Table 1

For each of the following types, write a problem of that type, draw a strip diagram or number bond (as appropriate), and formulate *two* equations for the problem, one using addition and one using subtraction.

1. A Put Together/Take Apart, Addend Unknown problem.

2. Two versions of a Compare, Difference Unknown problem: one formulated with "more" and one with "fewer."

3. Two versions of a Compare, Bigger Unknown problem: one formulated with "more" and one with "fewer."

4. Two versions of a Compare, Smaller Unknown problem: one formulated with "more" and one with "fewer."

SECTION 3.1 CLASS ACTIVITY 3-D

Identifying Problem Types and Difficult Language

CCSS CCSS SMP4, SMP7, 1.OA.1, 2.OA.1, Table 1

Do the following for each problem:

- Identify the type and subtype of the problem.
- Formulate an equation that fits with the language of the story problem, a "situation equation."
- Identify the language that might cause students to solve the problem incorrectly if they rely only on keywords.
- Make a math drawing for the problem.

1. Clare had 3 bears. After she got some more bears, Clare had 12 bears. How many bears did Clare get?

2. Clare has 12 bears altogether; 3 of the bears are red and the others are blue. How many blue bears does Clare have?

3. Kwon had some bugs. After he got 3 more bugs, Kwon had 12 bugs altogether. How many bugs did Kwon have at first?

4. Kwon has 12 red bugs. He has 3 more red bugs than blue bugs. How many blue bugs does Kwon have?

5. Nemili has 12 red triangles and 3 blue triangles. How many more red triangles does Nemili have than blue triangles?

6. Matt had some dinosaurs. After he gave away 5 dinosaurs, he had 9 dinosaurs left. How many dinosaurs did Matt have at first?

7. Matt has 5 fewer dinosaurs than bears. Matt has 9 dinosaurs. How many bears does Matt have?

SECTION 3.2 **CLASS ACTIVITY 3-E**

Mental Math

Try to find ways to make the problems that follow easy to do *mentally*. In each case, explain your method.

1. 7999 + 857 + 1

2. 367 + 98 + 2

3. 153 + 19 + 7

4. 7.89 + 6.95 + 0.05

SECTION 3.2 **CLASS ACTIVITY 3-F** 🏺

Children's Learning Paths for Single-Digit Addition

CCSS CCSS SMP7, K.OA, 1.OA.3, 1.OA.6, 2.OA.2

For children to develop fluency with the basic addition facts (from $1 + 1 = 2$ up to $9 + 9 = 18$), they first need extensive experience solving these basic addition problems in ways that make sense to them and that become increasingly sophisticated. The following levels describe the increasingly sophisticated methods children learn. (See [Car99], [CCOA11], and [Fus03].)

Level 1: Direct modeling, count all To add $5 + 4$, a child at this level counts out 5 things (or fingers), counts out another 4 things (or fingers), and then counts the total number of things (or fingers).

Level 2: Count on To add $6 + 3$, a child at this level imagines 6 things, says "six" (possibly elongating it, as in "siiiiix," while perhaps thinking about pointing along a collection of 6 things). Then the child says the next 3 number words, "seven, eight, nine," usually keeping track on fingers.

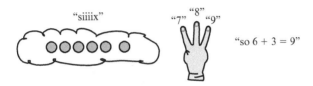

Count on from larger After children can count on from the first addend, they learn to count on from the larger addend. So to add $2 + 7$, a child at this level would count on from 7 instead of counting on from 2.

Level 3: Derived fact methods Children at this level use addition facts they already know to find related facts.

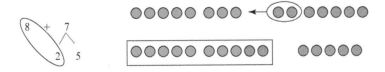

Make-a-ten method To calculate $8 + 7$ with this strategy, a child breaks 7 apart into $2 + 5$ so that a 10 can be made from the 8 and the 2. So the total is a 10 and 5 ones, which is 15 (as shown above).

Make-a-ten from larger The child makes a 10 with the larger number instead of just with the first addend.

Doubles ± 1 Children who know the doubles facts ($1 + 1$, $2 + 2$, up to $9 + 9$) can use these facts to find a related fact in which one of the addends is one more or one less than the addends in the double. For example, a child could determine that $6 + 7$ is 1 more than $6 + 6$.

1. For each method in the three levels, act out a few examples of how a child could use that method to solve a basic addition problem (adding two 1-digit numbers). For example, try 3 + 8 at levels 2 and 3 and try 9 + 4 and 7 + 5 at level 3.

2. What property of addition does the count on from larger method rely on? Explain, writing equations to demonstrate how the property is used.

3. Why is the property of addition you discussed in the previous part especially important for lightening the load of learning the table of basic addition facts (from 1 + 1 up to 9 + 9)?

4. What property of addition does the make-a-ten method rely on? Explain, writing equations to demonstrate how the property is used.

5. Why is the make-a-ten method especially important?

6. What property of addition does the doubles +1 method rely on? Explain, writing equations to demonstrate how the property is used.

7. Explain why the following three items are prerequisites for children to be able to understand and use the make-a-ten method fluently:

   a. For each counting number from 1 to 9, know the number to add to it to make 10—the partner to 10.

   b. For each counting number from 11 to 19, know that the number is a 10 and some ones. For example, know that 10 + 3 = 13 without counting and know that 13 decomposes as 10 + 3.

   c. For each counting number from 2 to 9, know all the ways to decompose it as a sum of two counting numbers (and know all the basic addition facts with sums up to 9).

SECTION 3.2 CLASS ACTIVITY 3-G 🏺

Children's Learning Paths for Single-Digit Subtraction

CCSS CCSS SMP7, K.OA.1, K.OA.2, 1.OA.4, 1.OA.6, 2.OA.2

For each of the subtraction methods listed below, act out a few examples of how a child could use that method to solve a basic subtraction problem.

Level 1: Direct modeling, take from To subtract 9 − 4, a child at this level counts out 9 things (or fingers), takes 4 things from 9 (or puts down 4 fingers), and then counts the number of things (or fingers) remaining.

Level 2: Count on to find the unknown addend The child views a subtraction problem as an unknown addend problem and counts on from the known addend to the total. So 13 − 9 = ? becomes 9 + ? = 13. The child counts on from 9, using fingers to keep track of how many have been counted on, and stops when the total is said. (This method is easier than counting down, which is difficult for most children. It makes subtraction as easy as addition.)

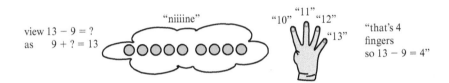

view 13 − 9 = ?
as   9 + ? = 13

"niiiine"

"10" "11" "12" "13"

"that's 4 fingers so 13 − 9 = 4"

Level 3: Make-a-ten methods

Make-a-ten with the unknown addend 14 − 8 = ? becomes 8 + ? = 14. The child figures that adding 2 to 8 makes 10, then adding another 4 makes 14, so the unknown addend is 2 + 4 = 6.

Subtract from ten This is like the previous method but the child breaks 14 into 10 and 4 and subtracts 8 from 10, leaving 2, then combines this 2 with the remaining 4 from 14 to make 6.

Subtract down to ten first To solve 12 − 3 = ? the child breaks 3 into 2 + 1, takes 2 from 12 to get down to 10, then takes the remaining 1 from 10 to get 9.

Compare and contrast the methods at level 3. When are the first two methods easier than the third? When is the third method easier than the first two?

SECTION 3.2 **CLASS ACTIVITY 3-H**

Reasoning to Add and Subtract

CCSS CCSS SMP2

1. John and Anne want to calculate $253 - 99$ by first calculating

$$253 - 100 = 153$$

John says that they must now *subtract* 1 from 153, but Anne says that they must *add* 1 to 153.

   a. Draw a number line (which need not be perfectly to scale) to help you explain who is right and why. Do not just say which answer is numerically correct; use the number line to help you explain why the answer must be correct.

   b. Explain in another way who is right and why.

2. Dakota says they are thinking of $253 - 99$ as how much they need to add to 99 to make 253. With Dakota's view of $253 - 99$ as an unknown addend or a difference in a comparison, what reasoning might they use to calculate the answer?

3. Jamarez says that he can calculate $253 - 99$ by adding 1 to both numbers and calculating $254 - 100$ instead.

   a. Draw a number line (which need not be perfectly to scale) to help you explain why Jamarez's method is valid.

   b. Explain in another way why Jamarez's method is valid.

    c. Could you adapt Jamarez's method to other subtraction problems, such as to the problem 324 − 298? Explain, and give several other examples.

4. Find ways to solve the addition and subtraction problems that follow *other than* by using the standard addition or subtraction algorithms. In each case, explain your reasoning, and—except for part (g)—write equations that correspond to your line of reasoning.

    a. 183 + 99

    b. 268 + 52

    c. 600 − 199

    d. 164 − 70

    e. 999 + 9999

    f. $10.00 − $2.99

    g. 2.99   (No equations are needed.)
       3.99
       1.99
    +4.99

Adding and Subtracting with Base-Ten Math Drawings

CCSS   CCSS SMP3, SMP5, 1.NBT.4

The hypothetical student work below is similar to actual first graders' work. These urban Latinx first-graders performed substantially above other first-graders and older children. (See [Fus97].)

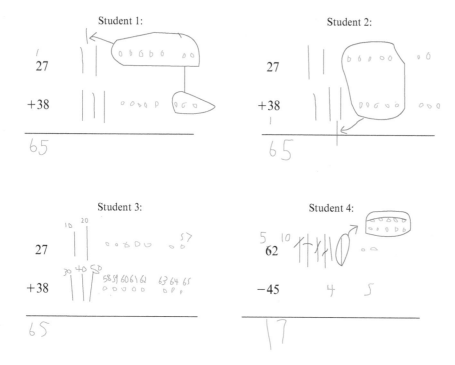

1. Examine and discuss the students' work. Compare the work of students 1, 2, and 3. In particular, compare the methods of student 1 and student 2 for adding 7 + 8. What mental method for subtracting 12 − 5 does the work of student 4 suggest?

2. Show how students 1, 2, and 3 might solve the addition problem 36 + 27 and how student 4 might solve the subtraction problem 43 − 18.

SECTION 3.3 CLASS ACTIVITY 3-J 🐛

Understanding the Standard Addition Algorithm

CCSS CCSS SMP3, SMP5, 2.NBT.7, 2.NBT.9

Materials Bundled toothpicks or base-ten blocks would be useful for this activity.

1. Add the numbers, using the standard paper-and-pencil method. Notice that regrouping is involved.

$$\begin{array}{r} 147 \\ + \ 195 \\ \hline \end{array}$$

2. If available, use bundled things to solve the addition problem 147 + 195. Make base-ten math drawings to indicate the process you used.

3. If available, use bundled things to solve the addition problem 147 + 195 *in a way that corresponds directly to the addition algorithm you used in part 1*. This may be different from what you did in part 2. Make base-ten math drawings to indicate the process you used. Compare with part 2.

4. Use expanded forms to solve the addition problem 147 + 195. First add like terms; then rewrite (regroup) the resulting number so that it is the expanded form of a number. This rewriting is the regrouping process.

$$\begin{array}{r} 1(100) + 4(10) + 7(1) \\ + \ 1(100) + 9(10) + 5(1) \\ \hline \end{array}$$

   $\longleftarrow$ First add like terms, remaining in expanded form.

   $\longleftarrow$ Then regroup so that you have the expanded form of a decimal number. You might want to take several steps to do so.

5. Compare and contrast your work in parts 1–4.

SECTION 3.3   CLASS ACTIVITY 3-K 🗑

Understanding the Standard Subtraction Algorithm

CCSS   CCSS SMP3, SMP5, 2.NBT.7, 2.NBT.9

Materials Bundled toothpicks or base-ten blocks would be useful for this activity.

1. Subtract the following numbers, using the standard paper-and-pencil method. Notice that regrouping is required.

$$\begin{array}{r} 125 \\ -68 \\ \hline \end{array}$$

2. If available, use bundled things to solve the subtraction problem $125 - 68$. Make base-ten math drawings to indicate the process.

3. If available, use bundled things to solve the subtraction problem $125 - 68$ *in a way that corresponds directly to the subtraction algorithm you used in part 1.* This may be different from what you did in part 2. Make base-ten math drawings to indicate the process. Compare with part 2.

4. Solve the subtraction problem $125 - 68$, but now use expanded forms. Start by rewriting the number 125 in expanded form. Rewrite the number in several steps, so that it will be easy to take 68 from 125. This rewriting is the regrouping process.

$$125 = 1(100) + 2(10) + 5(1) =$$

Write your regrouped number here $\longrightarrow$

Subtract 68: $-\, [6(10) + 8(1)]$

5. Compare and contrast your work in parts 1–4.

6. Use the standard algorithm to solve the following subtraction problem. Notice that regrouping across 0 is required.

$$\begin{array}{r} 104 \\ -47 \\ \hline \end{array}$$

Now explain the process in terms of bundled things or base-ten drawings.

How else could you solve the subtraction problem?

SECTION 3.3 CLASS ACTIVITY 3-L

A Third-Grader's Method of Subtraction

CCSS CCSS SMP3

When asked to compute $423 - 157$, Pat (a third-grader) wrote the following:

4−

30−

34−

300

266

"You can't take 7 from 3; it's 4 too many, so that's negative 4. You can't take 50 from 20; it's 30 too many, so that's negative 30; and with the other 4, it's negative 34. 400 minus 100 is 300, and then you take the 34 away from the 300, so it's 266."[1]

1. Discuss Pat's idea for calculating $423 - 157$. Is her method legitimate? Analyze Pat's method in terms of expanded forms.

2. Could you use Pat's idea to calculate $317 - 289$? If so, write what you think Pat might write, and also use expanded forms.

---

[1]This is taken from [Bri99, p. 263]

Regrouping in Base 12

CCSS CCSS SMP3

We can use the regrouping idea when objects are bundled in groups of a dozen instead of in groups of ten, as in the following problem.

> A store owner buys small, novelty party favors in bags of 1 dozen and boxes of 1-dozen bags (for a total of 144 favors in a box). The store owner has 5 boxes, 4 bags, and 3 individual party favors at the start of the month. At the end of the month, the store owner has 2 boxes, 9 bags, and 7 individual party favors left. How many favors did the store owner sell? Give the answer in terms of boxes, bags, and individual favors.

We can use a sort of *expanded form* for these party favors:

$$5(\text{boxes}) + 4(\text{bags}) + 3(\text{individual})$$
$$2(\text{boxes}) + 9(\text{bags}) + 7(\text{individual})$$

Solve this problem by regrouping among the boxes, bags, and individual party favors.

SECTION 3.3 **CLASS ACTIVITY 3-N**

Regrouping in Base 60

CCSS CCSS SMP3

We can use the regrouping idea with time. Just as 1 hundred is 10 tens and 1 ten is 10 ones, 1 hour is 60 minutes and 1 minute is 60 seconds. The following problem asks you to regroup among hours, minutes, and seconds:

> Ruth runs around a lake two times. The first time takes 1 hour, 43 minutes, and 38 seconds. The second time takes 1 hour, 48 minutes, and 29 seconds. What is Ruth's total time for the two laps? Give the answer in hours, minutes, and seconds.

We can use a sort of *expanded form* for time:

$$1(\text{hour}) + 43(\text{minutes}) + 38(\text{seconds})$$
$$1(\text{hour}) + 48(\text{minutes}) + 29(\text{seconds})$$

Solve this problem by regrouping among hours, minutes, and seconds.

SECTION 3.4 | **CLASS ACTIVITY 3-O** 🏺

Why Do We Add and Subtract Fractions the Way We Do?

CCSS CCSS SMP3, 4.NF.3a, 5.NF.1, 5.NF.2

Materials For part 2 of this activity, each person will need at least five identical strips of paper (or card stock).

1. When two fractions have the same denominator, we add or subtract them by keeping the same denominator and adding or subtracting the numerators. For example,

$$\frac{1}{5} + \frac{3}{5} = \frac{1+3}{5} = \frac{4}{5}$$

Patti says: "We should add the tops *and* the bottoms." She shows you this picture to explain why:

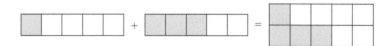

So according to Patti:

$$\frac{1}{5} + \frac{3}{5} = \frac{1+3}{5+5} = \frac{4}{10}$$

  a. Why is Patti's method *not* a valid way to add fractions, and why doesn't Patti's picture prove that fractions can be added in her way? Critique Patti's reasoning.

  b. How could Patti use estimation and benchmark fractions to see if her answer is reasonable?

  c. Explain why the proper way to add $\frac{1}{5} + \frac{3}{5}$ and $\frac{2}{5} + \frac{4}{5}$ makes sense, using our definition of fractions.

2. You need five identical strips of paper for this part.

   a. Label one of the strips "1 whole," and fold and label two other strips as indicated:

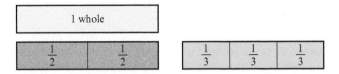

   b. Fold and place your halves and thirds strips to show these lengths:

$$\frac{1}{2} + \frac{1}{3}, \qquad \frac{2}{3} - \frac{1}{2}, \qquad \frac{2}{3} + \frac{1}{2}$$

   Make drawings to record your work.

   c. Discuss: Your folded strips in part b show the requested lengths, so why are you not done solving the problem?

   d. Fold the remaining two strips, one in halves, the other in thirds. Then fold each of them again into parts that will be better for expressing the lengths in part b. Relate this to the numerical procedure for adding and subtracting fractions.

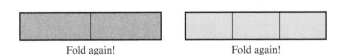

Fold again!   Fold again!

Critiquing Mixed Number Addition and Subtraction Methods

CCSS CCSS SMP3, 5.NF.1

1. Each of the problems that follows shows some student work. Discuss the work: What is correct and what is not correct? In each case, either complete the work or modify it to make it correct. *Use the student's work; do not start from scratch.*

   a. Subtract: $3\frac{1}{4} - 1\frac{3}{4}$.

   student 1:
   $$\begin{array}{r} 2 \\ \cancel{3}\frac{\overset{10}{\cancel{4}}1}{4} \\ - 1\frac{3}{4} \\ \hline 1\frac{7}{4} \end{array}$$

   student 2:
   $$\begin{array}{r} 2\frac{4}{4} \\ \cancel{3}\frac{1}{4} \\ - 1\frac{3}{4} \\ \hline \end{array} \Big\} \frac{1}{4}$$

   b. There are $2\frac{1}{3}$ cups of milk in a bowl. How much milk must be added to the bowl so that there will be 3 cups of milk in the bowl?

   $$2 \qquad \underbrace{2\frac{1}{3} \quad 2\frac{2}{3} \quad 2\frac{3}{3}}_{\text{2 more}} \overset{3}{\phantom{2\frac{3}{3}}}$$

   c. There were 5 pounds of apples in a bag. After some of the apples were removed from the bag, there were $3\frac{1}{4}$ pounds of apples left. How many pounds of apples were removed?

   $$5 \qquad \frac{4}{4} \quad \frac{4}{4} \quad \frac{4}{4} \quad \frac{4}{4} \quad \cancel{\frac{4}{4}}$$
   $$\frac{1}{4} \quad \frac{3}{4}$$

   d. Add: $2\frac{2}{3} + 1\frac{2}{3}$.

   $2\frac{2}{3} + 1\frac{2}{3} = 3\frac{4}{6}$ because the fraction part is 4 out of 6.

2. Find at least two different ways to calculate

$$7\frac{1}{3} - 4\frac{1}{2}$$

and to give the answer as a mixed number. In each case, explain why your method makes sense.

SECTION 3.4 CLASS ACTIVITY 3-Q 🍎

Are These Word Problems for $\frac{1}{2} + \frac{1}{3}$?

CCSS CCSS SMP4, 5.NF.2

For each problem, determine if it is a problem for $\frac{1}{2} + \frac{1}{3}$. If not, explain why not, and explain how to solve the problem if possible. If there is not enough information to solve the problem, explain why not.

1. Tom pours $\frac{1}{2}$ cup of water into an empty bowl and then pours $\frac{1}{3}$ cup of water into the bowl. How many cups of water are in the bowl now?

2. Tom pours $\frac{1}{2}$ cup of water into an empty bowl and then pours in another $\frac{1}{3}$. How many cups of water are in the bowl now?

3. $\frac{1}{2}$ of the land in Heeltoe County is covered with forest; $\frac{1}{3}$ of the land in the adjacent Toejoint County is covered with forest. What fraction of the land in the two-county Heeltoe–Toejoint region is covered with forest?

4. $\frac{1}{2}$ of the land in Heeltoe County is covered with forest; $\frac{1}{3}$ of the land in the adjacent Toejoint County is covered with forest. Heeltoe and Toejoint counties have the same land area. What fraction of the land in the two-county Heeltoe–Toejoint region is covered with forest?

5. Students at Martin Luther King Elementary School could vote for all the lunch choices they like. $\frac{1}{2}$ of the children say they like to have pizza for lunch; $\frac{1}{3}$ of the children say they like to have a hamburger for lunch. What fraction of the children at Martin Luther King Elementary School would like to have either pizza or a hamburger for lunch?

6. $\frac{1}{2}$ of the children at Timothy Elementary School like to have pizza for lunch, and the other half does not like to have pizza for lunch. Of the children who do not like to have pizza for lunch, $\frac{1}{3}$ like to have a hamburger for lunch. What fraction of the children at Timothy Elementary School like to have either pizza or a hamburger for lunch?

SECTION 3.4   CLASS ACTIVITY 3-R

Are These Word Problems for $\frac{1}{2} - \frac{1}{3}$?

CCSS   CCSS SMP4, 5.NF.2

For each of the following problems, determine if it can be solved by subtracting $\frac{1}{2} - \frac{1}{3}$. If not, explain why not, and explain how the problem can be solved if there is enough information to do so. If there is not enough information to solve the problem, explain why not.

1. Zelha pours $\frac{1}{2}$ cup of water into an empty bowl and then pours out $\frac{1}{3}$. How much water is in the bowl now?

2. Zelha pours $\frac{1}{2}$ cup of water into an empty bowl and then pours out $\frac{1}{3}$ cup of water. How much water is in the bowl now?

3. Zelha pours $\frac{1}{2}$ cup of water into an empty bowl and then pours out $\frac{1}{3}$ of the water that is in the bowl. How much water is in the bowl now?

4. Yesterday James ate $\frac{1}{2}$ of a pizza, and today he ate $\frac{1}{3}$ of a pizza of the same size. How much more pizza did James eat yesterday than today?

5. Yesterday James ate $\frac{1}{2}$ of a pizza, and today he ate $\frac{1}{3}$ of the whole pizza. Nobody else ate any of that pizza. How much pizza is left?

6. Yesterday James ate $\frac{1}{2}$ of a pizza, and today he ate $\frac{1}{3}$ of the pizza that was left over from yesterday. Nobody else ate any of that pizza. How much pizza is left?

SECTION 3.4 **CLASS ACTIVITY 3-S**

What Fraction Is Shaded?

CCSS CCSS SMP1, 5.NF.2

Materials Download 3-1 at bit.ly/2SWWFUX has copies of the squares for experimenting.

For each square shown, determine the fraction of the area of the square that is shaded. Explain your reasoning. You may assume that all lengths that appear to be equal really are equal. Try not to use any area formulas. Apply your knowledge of how to add and subtract fractions!

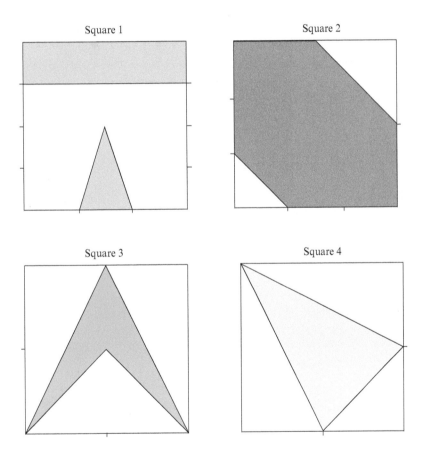

Square 1

Square 2

Square 3

Square 4

Addition with Whole Numbers, Decimals, Fractions, and Mixed Numbers: What Are Common Ideas?

CCSS   CCSS SMP8

Think about how we add (and subtract) whole numbers, decimals, fractions, and mixed numbers, and think about the ideas and reasoning that are involved. What is common in the way we add (and subtract) across all these different kinds of numbers?

Using Word Problems to Find Rules for Adding and Subtracting Negative Numbers

CCSS   CCSS SMP2, SMP8, 7.NS.1

This activity relies on some of the different types of addition and subtraction word problems discussed in Section 3.1. You will write word problems and use these problems to see why some of the rules for adding and subtracting with negative numbers make sense.

In writing your word problems, consider that negative numbers are nicely interpreted as temperatures below zero, locations below ground (as in a building that has basement stories, for example), locations below sea level, amounts owed, or negatively charged particles.

1. a. Write and solve an Add To or Put Together/Take Apart problem for $(-5) + 5 = ?$.

    b. Think more generally about your problem in part (a) and about its solution, and imagine changing the numbers. What can you conclude about $(-N) + N$? Write an equation that shows your conclusion.

2. a. Write and solve a Take From problem for $(-2) - 5 = ?$.

b. Write and solve a Compare problem for $(-2) - 5 = ?$ in which one quantity is -2 and the other quantity is 5 less and is unknown.

c. Think more generally about your problems in parts (a) and (b) and about their solutions, and imagine changing the numbers. What can you conclude about how $(-A) - B$ and $A + B$ are related? Write an equation that shows your conclusion.

3. a. Write and solve a Compare problem for $2 - (-5) = ?$ in which one quantity is 2, the other quantity is -5, and the difference between the two quantities is unknown.

b. Think more generally about your problem in part (a) and about its solution, and imagine changing the numbers. What can you conclude about $A - (-B)$? Write an equation that shows your conclusion.

SECTION 4.1   CLASS ACTIVITY 4-A

What Is Multiplication?

1. What are your initial thoughts on how you might explain what multiplication means? For example, what does 3×5 mean?

2. Write a simple word problem you might use to illustrate what 3×5 means. Make a math drawing for your word problem.

3. Identify the referent units in your word problem in part 2, and its solution:

    The 3 is 3 of what?

    The 5 is 5 of what?

    The 15 is 15 of what?

4. Compare the answers you wrote for parts 1, 2, and 3 with those of several classmates. What are similarities? What are differences?

5. Would the meaning you described in part 1 apply to $\frac{1}{2} \cdot \frac{1}{3}$? Why or why not?

SECTION 4.1 CLASS ACTIVITY 4-B 🏛

How Can We Show Multiplicative Structure?

CCSS CCSS SMP2, SMP6, SMP7, 3.OA.3, 4.OA.1, 4.OA.2, 7.RP.8b

In this Class Activity you will use our definition of multiplication to explain why we can multiply to solve a problem. Here is an example:

> *Pencil Problem:* A teacher has 3 boxes of pencils. Each box contains 12 pencils. How many pencils does the teacher have?

Let 1 group = 12 pencils, the number of pencils in 1 box. So we can rephrase the problem as asking: 3 groups of 12 pencils are how many pencils? According to our definition of multiplication, this number of pencils is $3 \cdot 12$.

1 box	1 box	1 box
12 pencils	12 pencils	12 pencils

3 groups of 12 pencils are how many pencils?
$$3 \quad \cdot \quad 12 \quad = \quad ?$$
groups   pencils   pencils

1. For each of the following, use our definition of multiplication to explain why the problem can be solved by multiplying. As in the example above, make a math drawing and write and annotate a multiplication equation to model the problem, using a question mark for the unknown.

   a. How many ladybugs are in the array below?

   b. Eva's puppy weighs 9 pounds. Micah's puppy weighs 4 times as much as Eva's puppy. How much does Micah's puppy weigh?

c. Fran has 3 pairs of pants (pants 1, 2, and 3) that coordinate perfectly with 4 different shirts (shirts A, B, C, and D). How many different outfits consisting of a pair of pants and a shirt can Fran make from these clothes?

d. If 1 meter of a rope weighs 9 grams, how much do 7 meters of the same kind of rope weigh?

e. If a piece of rope is 4 feet long and weighs 1 pound, then how long is a piece of rope of the same type that weighs 7 pounds?

2. You have 6 pairs of gloves, one pair in each of these colors: red, blue, yellow, green, orange, and purple. Each pair consists of a left glove and a right glove.

a. Write a $2 \cdot 6 = ?$ word problem that uses this context.

b. Write a $6 \cdot 6 = ?$ word problem that uses this context.

SECTION 4.1 **CLASS ACTIVITY 4-C**

Writing Multiplication Word Problems

CCSS CCSS SMP2, SMP6

Recall our definition of multiplication.

$$\underset{\text{groups}}{\text{M}} \quad \bullet \quad \underset{\text{units}}{\text{N}} \quad = \quad \underset{\text{units}}{\text{P}}$$

M groups of N units are P units

1. Write a simple word problem for $4 \cdot 8 = ?$ Explain why the problem can be solved by multiplying $4 \cdot 8$ (make a math drawing to aid your explanation, if possible).

2. Write an Array problem for $3 \cdot 7 = ?$ Explain why the problem can be solved by multiplying $3 \cdot 7$. How else can the problem be solved? Explain.

3. Write a Multiplicative Comparison problem for $5 \cdot 9 = ?$ Draw a strip diagram for the problem and explain why the problem can be solved by multiplying $5 \cdot 9$.

4. Write an Ordered Pair problem for $8 \cdot 5 = ?$ Explain why the problem can be solved by multiplying $8 \cdot 5$. How else can the problem be solved? Explain.

5. Write a multiplication word problem that concerns the volume (e.g., in cups, milliliters, or liters) and weight (e.g., in pounds, ounces, grams, or kilograms) of some kind of food. Use our definition to explain why it is a multiplication problem.

SECTION 4.2 | CLASS ACTIVITY 4-D 🍎

What Is Special about Multiplying by 10?

CCSS CCSS SMP3, 4.NBT.1, 5.NBT.1, 5.NBT.2

1. Do you know a quick way to multiply by 10 or 100 or 1000 mentally? For example, what are 10×47 and 100×47 and 1000×47?

2. Which of the following statements are correct and appropriate to use with all numbers in base ten?

   a. To multiply a number by 10, put a 0 at the end of the number.

   b. To multiply a number by 10, move the decimal point one place to the right.

   c. To multiply a number by 10, move all the digits one place to the left.

3. Explain why statement 2(c) about multiplying by 10 is true. To do so, you can use the math drawing below, which represents 10×23, or make your own math drawing. What happens to each of the 2 tens and what happens to each of the 3 ones when we multiply 23 by 10?

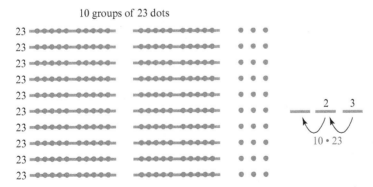

10 groups of 23 dots

SECTION 4.3 **CLASS ACTIVITY 4-E**

Why Is the Commutative Property of Multiplication *Not* Obvious and Why Is It True?

CCSS CCSS SMP7, 3.OA.5

1. Write one word problem for $7 \cdot 4$ and another for $4 \cdot 7$ (using our definition of multiplication). Before determining the answers to the problems, discuss whether it would necessarily be obvious to a student that the answers are the same.

2. View the following array in two ways to explain why $4 \cdot 7 = 7 \cdot 4$.

3. Use a math drawing and our definition of multiplication to explain why

$$3 \cdot 6 = 6 \cdot 3$$

without computing that both products are 18. Use the idea of seeing the same thing in two ways!

4. If A and B are counting numbers, why must $A \cdot B$ necessarily be equal to $B \cdot A$? How do we know this even if we don't know what the products $A \cdot B$ and $B \cdot A$ are?

Why Can We Multiply to Find the Area of a Rectangle?

CCSS CCSS SMP7, 3.MD.7a

1. The large rectangle shown here is 9 centimeters wide and 5 centimeters tall. Use the definition of multiplication to explain why you can find the area of the rectangle by multiplying.

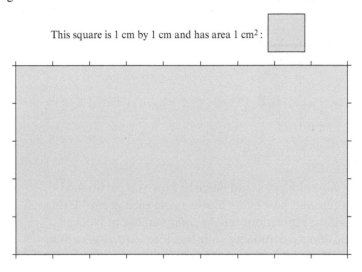

This square is 1 cm by 1 cm and has area 1 cm² :

2. Using the rectangle in part 1, explain why

$$5 \cdot 9 = 9 \cdot 5$$

3. How can you use area to explain why the commutative property is true more generally? Why is it true that

$$A \cdot B = B \cdot A$$

no matter what the counting numbers A and B are?

SECTION 4.3 **CLASS ACTIVITY 4-G**

Why Can We Multiply to Find the Volume of a Box?

CCSS CCSS SMP7, 5.MD.5a

Materials You will need a set of blocks for parts 1 and 2 of this activity.

1. If you have cubic-inch blocks available, build a box that is 3 inches wide, 2 inches long, and 4 inches tall. It should look like this:

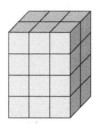

2. Partition your box into natural groups of blocks, and describe how you partitioned the box. How many groups were there and how many blocks made each group? Using multiplication, write the corresponding expressions for the total number of blocks that make the box. Now repeat, this time partitioning your box into natural groups in a different way.

3. The figures below show different ways of partitioning a box into groups. In each case, describe the number of groups and the number of blocks that make each group. Then write an expression for the total number of blocks. Your expressions should use multiplication, parentheses, and the numbers 2, 3, and 4 only.

A.

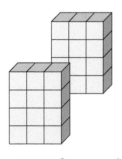

_____ groups of _____ blocks

Write an expression using multiplication, parentheses, and the numbers 2, 3, and 4 for the total number of blocks:

_____

B.

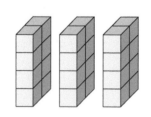

_____ groups of _____ blocks

Write an expression using multiplication, parentheses, and the numbers 2, 3, and 4 for the total number of blocks:

_____

C.

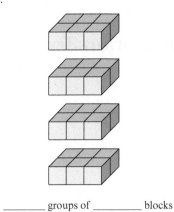

_____ groups of _____ blocks

Write an expression using multiplication, parentheses, and the numbers 2, 3, and 4 for the total number of blocks:

_____

D.

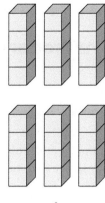

_____ groups of _____ blocks

Write an expression using multiplication, parentheses, and the numbers 2, 3, and 4 for the total number of blocks:

_____

E.

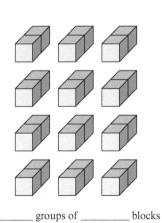

_____ groups of _____ blocks

Write an expression using multiplication, parentheses, and the numbers 2, 3, and 4 for the total number of blocks:

_____

F.

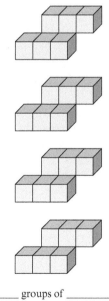

_____ groups of _____ blocks

Write an expression using multiplication, parentheses, and the numbers 2, 3, and 4 for the total number of blocks:

_____

SECTION 4.3   CLASS ACTIVITY 4-H

Why Is the Associative Property of Multiplication True?

CCSS   CCSS SMP7, 3.OA.5

Gum Problem: You have 4 bags of gum. Each bag has 8 packs. Each pack has 5 sticks of gum. How many sticks of gum do you have?

1. Find two different ways to solve the Gum Problem by multiplying. Explain your reasoning in each case. Is one way easier than the other?

2. Explain how to interpret $4 \cdot (8 \cdot 5)$ and $(4 \cdot 8) \cdot 5$ in terms of the Gum Problem. Use our definition of multiplication in each case.

3. Based on the situation of the Gum Problem, how do we know that the following equation *must necessarily* be true even if we didn't know how many sticks of gum there are?

$$4 \cdot (8 \cdot 5) = (4 \cdot 8) \cdot 5$$

Smiley-Face Stickers: You have 5 bags. Each bag has 4 sheets. Each sheet has 12 smiley-face stickers.

4. Use the context of the Smiley-Face Stickers and our definition of multiplication to explain how we can tell that the following equation must necessarily be true *without figuring out how many stickers there are.*

$$5 \cdot (4 \cdot 12) = (5 \cdot 4) \cdot 12$$

5. Explain how to calculate the number of smiley-face stickers in two different ways. Is one way easier than the other?

6. Use some of the figures and the expressions you wrote for part 3 of Class Activity 4-G to help you explain why

$$(A \times B) \times C = A \times (B \times C)$$

for counting numbers A, B, C.

SECTION 4.3 CLASS ACTIVITY 4-I 🍎

How Can We Use the Associative and Commutative Properties of Multiplication?

CCSS CCSS SMP2, SMP7, 3.NBT.3

1. You have 7 bags of gum. Each bag contains 6 packs. Each pack contains 5 pieces of gum. Explain how to interpret

$$(7 \cdot 6) \cdot 5 \quad \text{and} \quad 7 \cdot (6 \cdot 5)$$

in terms of the gum. Why must they be equal? Which of the two is easier to calculate?

2. **a.** To calculate $4 \cdot 60$ mentally, Zeph figures $4 \cdot 6 = 24$ and then puts a 0 at the end to make 240. Does Zeph's method give the right answer? Will Zeph's method work for other multiplication problems?

b. Use the equations and math drawing below to explain why Zeph's method works. Which property of multiplication does Zeph's method rely on?

$$
\begin{aligned}
4 \cdot 60 &= 4 \cdot (6 \cdot 10) \\
&= (4 \cdot 6) \cdot 10 \\
&= 24 \cdot 10 \\
&= 240
\end{aligned}
$$

3. Explain how to use the associative property of multiplication to make $28 \cdot 0.25$ easy to calculate mentally. Write equations to show how the associative property is used. How else can you think about the problem?

4. How is the associative property of multiplication involved when viewing 20 groups of 10 as 2 groups of 100? Write equations to help explain.

5. There are 21 bags with 2 marbles in each bag. Ben calculates the number of marbles there are in all by counting by twos 21 times:

$$2, 4, 6, 8, 10, \ldots, 40, 42$$

Kaia calculates $21 + 21 = 42$ instead.

Discuss the two calculation methods. Are both legitimate? How could you relate the two methods? Is this related to a property of multiplication? Explain.

SECTION 4.4 CLASS ACTIVITY 4-J 🕱

Why Is the Distributive Property True?

CCSS   CCSS SMP7, 3.MD.7c

1. There are 6 goodie bags. Each goodie bag contains 3 eraser tops and 4 stickers.

Explain why each of the two expressions

$$6 \cdot 3 + 6 \cdot 4 \quad \text{and} \quad 6 \cdot (3 + 4)$$

describes the total number of items in the goodie bags and therefore why

$$6 \cdot (3 + 4) = 6 \cdot 3 + 6 \cdot 4$$

Your explanation should *not* involve any calculations.

Then discuss how to view your explanation as explaining why the distributive property is true for all counting numbers.

2. Use the different shading shown in the rectangle, and use our definition of multiplication, to explain why

$$3 \cdot (2 + 4) = 3 \cdot 2 + 3 \cdot 4$$

Your explanation should be general in the sense that you could use it to explain why

$$A \cdot (B + C) = A \cdot B + A \cdot C$$

for *all* counting numbers *A, B,* and *C*.

3. Make a rough drawing of an array and shade it to illustrate the equation

$$8 \cdot (10 + 5) = 8 \cdot 10 + 8 \cdot 5$$

SECTION 4.4   CLASS ACTIVITY 4-K 🍎

Applying the Distributive Property to Calculate Flexibly

CCSS   CCSS SMP2, SMP7

1. Use the multiplication facts

$$15 \cdot 15 = 225$$
$$2 \cdot 15 = 30$$

to help you mentally calculate

$$17 \cdot 15$$

Explain how your calculation method is related to the array below, which consists of 17 rows of 15 dots. Also write equations showing how your mental strategy for calculating $17 \cdot 15$ involves the distributive property.

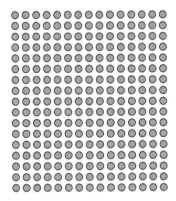

2. Write an equation that uses subtraction and the distributive property and that goes along with the following array:

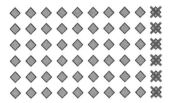

3. Mentally calculate $20 \cdot 15$, and use your answer to mentally calculate $19 \cdot 15$. Write an equation that uses subtraction and the distributive property and that goes along with your strategy. Without drawing all the detail, make a rough math drawing of an array that illustrates this calculation strategy.

Critique Multiplication Strategies

CCSS   CCSS SMP3

1. Kylie has an idea for how to calculate 23 · 23. She says,

Twenty times 20 is 400, and 3 times 3 is 9; so 23 · 23 should be 400 plus 9, which is 409.

Is Kylie's method valid? If not, how could you modify her work to make it correct? Don't just start over in a different way; work with Kylie's idea. Use the large square below, which consists of 23 rows with 23 small squares in each row, to help you explain your answer.

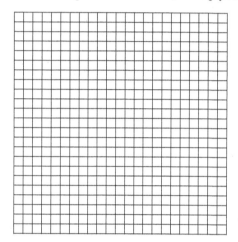

2. Annie is working on the multiplication problem 19 · 21. She says that 19 · 21 should equal 20 · 20 because 19 is one less than 20 and 21 is one more than 20.

This is a wonderful idea, but is Annie correct? If not, use the diagram below to help you explain why not. There are 20 rows of dots with 21 dots in each row.

3. Mary is working on the multiplication problem 19 · 21. She says that 19 · 21 is 21 less than 20 · 21, and 20 · 21 is 20 more than 20 · 20, which she knows is 400. Mary thinks this ought to help her calculate 19 · 21, but she can't quite figure it out.

Discuss Mary's idea in detail. Can you make her idea work?

Using Properties of Arithmetic to Aid the Learning of Basic Multiplication Facts

CCSS CCSS 3.OA.5

In school, students must learn the single-digit multiplication facts from $1 \times 1 = 1$ to $9 \times 9 = 81$. By learning relationships among the facts, students can structure their understanding of the single-digit facts in order to learn them better.

$2 \times 2 = 4$	3×2	4×2	5×2	6×2	7×2	8×2	9×2
$2 \times 3 = 6$	$3 \times 3 = 9$	4×3	5×3	6×3	7×3	8×3	9×3
$2 \times 4 = 8$	$3 \times 4 = 12$	$4 \times 4 = 16$	5×4	6×4	7×4	8×4	9×4
$2 \times 5 = 10$	$3 \times 5 = 15$	$4 \times 5 = 20$	$5 \times 5 = 25$	6×5	7×5	8×5	9×5
$2 \times 6 = 12$	$3 \times 6 = 18$	$4 \times 6 = 24$	$5 \times 6 = 30$	$6 \times 6 = 36$	7×6	8×6	9×6
$2 \times 7 = 14$	$3 \times 7 = 21$	$4 \times 7 = 28$	$5 \times 7 = 35$	$6 \times 7 = 42$	$7 \times 7 = 49$	8×7	9×7
$2 \times 8 = 16$	$3 \times 8 = 24$	$4 \times 8 = 32$	$5 \times 8 = 40$	$6 \times 8 = 48$	$7 \times 8 = 56$	$8 \times 8 = 64$	9×8
$2 \times 9 = 18$	$3 \times 9 = 27$	$4 \times 9 = 36$	$5 \times 9 = 45$	$6 \times 9 = 54$	$7 \times 9 = 63$	$8 \times 9 = 72$	$9 \times 9 = 81$

1. Examine the darkly colored, lightly colored, and uncolored regions in the above multiplication table. Explain how to obtain the uncolored facts quickly and easily from the shaded facts. In doing so, what property of arithmetic do you use? How does knowing this property of arithmetic lighten the load for students of learning the single-digit multiplication facts?

2. Multiplication facts involving the numbers 6, 7, and 8 are often hard to learn. For each fact in the lightly colored regions in the table, describe one or more ways to derive it from facts in the darkly colored region by applying properties of arithmetic. Use arrays and equations to show the reasoning, as in these examples for 3×7, which use the distributive property:

$$3 \times 7 = 2 \times 7 + 1 \times 7$$
$$= 14 + 7 = 21$$

$$3 \times 7 = 3 \times 3 + 3 \times 4$$
$$= 9 + 12 = 21$$

3. The *5× table* is easy to learn because it is "half of the *10× table.*"

$$5 \times 1 = 5 \qquad 10 \times 1 = 10$$
$$5 \times 2 = 10 \qquad 10 \times 2 = 20$$
$$5 \times 3 = 15 \qquad 10 \times 3 = 30$$
$$5 \times 4 = 20 \qquad 10 \times 4 = 40$$
$$5 \times 5 = 25 \qquad 10 \times 5 = 50$$
$$5 \times 6 = 30 \qquad 10 \times 6 = 60$$
$$5 \times 7 = 35 \qquad 10 \times 7 = 70$$
$$5 \times 8 = 40 \qquad 10 \times 8 = 80$$
$$5 \times 9 = 45 \qquad 10 \times 9 = 90$$

Write an equation showing the relationship between 10×7 and 5×7 that fits with the statement about the *5× table* being half of the *10× table*. Which property of arithmetic do you use?

4. The *9× table* is easy to learn because you can use subtraction, as shown next. Explain why this way of multiplying by 9 is valid.

$$9 \times 1 = 10 - 1 = 9$$
$$9 \times 2 = 20 - 2 = 18$$
$$9 \times 3 = 30 - 3 = 27$$
$$9 \times 4 = 40 - 4 = 36$$
$$9 \times 5 = 50 - 5 = 45$$
$$9 \times 6 = 60 - 6 = 54$$
$$9 \times 7 = 70 - 7 = 63$$
$$9 \times 8 = 80 - 8 = 72$$
$$9 \times 9 = 90 - 9 = 81$$

5. What is another pattern (other than the one described in part 4) in the *9× table*?

CLASS ACTIVITY 4-N

Solving Arithmetic Problems Mentally

CCSS CCSS SMP7

For each of the following arithmetic problems, describe a way to make the problem easy to solve mentally:

1. 4×99

2. 16×25 (Try to find several ways to solve this problem mentally.)

3. $45\% \times 680$

4. 12×125 (Try to find several ways to solve this problem mentally.)

5. $125\% \times 120$

SECTION 4.5   **CLASS ACTIVITY 4-O**

Writing Equations That Correspond to a Method of Calculation

CCSS   CCSS SMP7

Each arithmetic problem in this activity has a description of the problem solution. In each case, write a sequence of equations that corresponds to the given description. Which properties of arithmetic were used and where? Write your equations in the following form:

$$\text{original} = \text{some expression}$$
$$= \vdots$$
$$= \text{some expression}$$

1. What is 55% of 120?

   Half of 120 is 60. 10% of 120 is 12, so 5% of 120 is half of that 10%, which is 6. So the answer is 60 plus 6, which is 66.

2. What is 35% of 80?

   25% is $\frac{1}{4}$, so 25% of 80 is $\frac{1}{4}$ of 80, which is 20. 10% of 80 is 8. So 35% of 80 is 20 plus 8, which is 28.

3. What is 90% of 350?

   10% of 350 is 35. Taking 35 away from 350 leaves 315. So the answer is 315.

4. What is 12.5% of 1800?

   Half is 900 and half of that is 450, so that's 25%. Then half of that is 225 which is 12.5%.

CLASS ACTIVITY 4-P

Showing the Algebra in Mental Math

CCSS CCSS SMP7

For each arithmetic problem in this activity, find ways to use properties of arithmetic to make the problem easy to do mentally. Describe your method in words, and write equations that correspond to your method. Write your equations in the following form:

$$\begin{aligned} \text{original} &= \text{some expression} \\ &= \vdots \\ &= \text{some expression} \end{aligned}$$

1. 6×12 (Try to find several different ways to solve this problem mentally.)

2. $24 \cdot 25$ (Try to find several different ways to solve this problem mentally.)

3. $5\% \cdot 48$

4. $15\% \cdot \$44$

5. $26\% \cdot 840$

6. $9 \cdot 99$ (Try to find several different ways to solve this problem mentally.)

SECTION 4.6 CLASS ACTIVITY 4-Q

How Can We Develop and Understand the Standard Multiplication Algorithm?

CCSS CCSS SMP7, 4.NBT.5

Materials For experimenting, you might like to use Downloads 4-1, 4-2, 4-3 at bit.ly/2SWWFUX.

1. Imagine you are a fourth-grade student who is ready to learn about multiplying multi-digit numbers. Let's say you are fluent with the one-digit multiplications 1×1 through 9×9 and you understand multiplication with one-digit multiples of 10 such as 7×90 and 30×80. How could you use what you know to figure out how many dots are in the array below? The array has 6 rows and 38 columns.

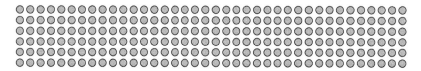

2. Use our definition of multiplication to explain why the number of dots in the array above can be found by multiplying 6×38. Then use the distributive property to calculate 6×38 in one or more ways. For each way, make a rough drawing to indicate how it is related to the array of dots. Are any of these like what you did in part 1?

3. Use the partial-products method to calculate 6×38. Does this method correspond to any of your methods in parts 1 or 2? If not, go back to parts 1 and 2 and look for corresponding methods.

4. Imagine once again that you are that fourth-grade student from part 1. How could you use what you know to figure out how many small squares are in the array below? The array has 23 rows and 45 columns.

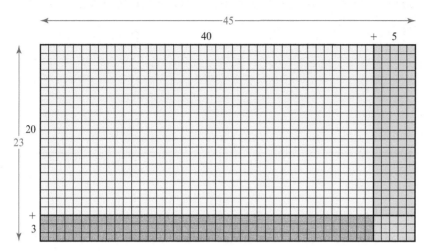

5. Use our definition of multiplication to explain why the number of small squares in the array above can be found by multiplying 23 × 45. Then use the distributive property to calculate 23 × 45 in one or more ways. For each way, make a rough drawing to indicate how it is related to the array of small squares. Are any of these like what you did in part 4?

6. Use the partial-products method to calculate 23 × 45. Does this method correspond to any of your methods in parts 4 or 5? If not, go back to parts 4 and 5 and look for corresponding methods. In part 5, decompose 23 and 45 into their place value parts.

7. Calculate 23 × 45 using the common method for writing the standard algorithm. Relate the two lines in the calculation to the array in part 4 and to a way to apply the distributive property to 23 × 45.

8. Make a rough sketch to indicate an array of small squares for 46 × 53. Your sketch should not show all the small squares and it does not have to be to scale. Use your sketch to explain why the partial-products method calculates the correct answer to 46 × 53. Begin your explanation by using the definition of multiplication to relate the array to the multiplication problem.

SECTION 5.1 CLASS ACTIVITY 5-A 🍎

Extending Multiplication to Fractions, Part I

CCSS CCSS SMP6, 4.NF.4

In this Class Activity, use our definitions of multiplication and of fractions throughout. Recall how our definition of multiplication works. For example:

$$3 \quad \bullet \quad \frac{3}{4} \quad = \quad ?$$

groups   units   units

3 groups of $\frac{3}{4}$ of a unit are how many units?

1. Write a simple word problem and make a math drawing for the example above.

2. If 1 serving of cereal is $\frac{2}{3}$ cups, then how many cups are in 5 servings of cereal? Make a math drawing and write and annotate an equation for this problem.

3. Interpret the equation below using our definition of multiplication. Then write a word problem and make and explain a math drawing for it.

$$4 \cdot \frac{3}{5} = ?$$

4. If you had 3 servings of a sports drink and one serving is $\frac{4}{5}$ liters, then how many liters did you have? Make a math drawing and write and annotate an equation for this problem.

5. One serving of vegetables is $\frac{1}{2}$ cup. You want to make $2\frac{1}{2}$ servings of vegetables. How many cups of vegetables is that? Make a math drawing and write and annotate an equation for this problem.

6. Interpret the equation below using our definition of multiplication. Then write a word problem and make and explain a math drawing for it.

$$1\frac{1}{2} \cdot \frac{2}{3} = ?$$

SECTION 5.1 **CLASS ACTIVITY 5-B**

Making Multiplicative Comparisons

CCSS CCSS SMP2, 4.OA.2

Button Strip A has 12 buttons, and Button Strip B has 72 buttons.

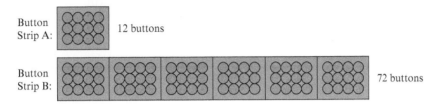

1. How many groups of Button Strip A does it take to make the exact same number of buttons as Button Strip B?

    How many groups of Button Strip B does it take to make the exact same number of buttons as Button Strip A?

2. Use your answers to part 1 to write multiplication equations that relate 12 and 72. Explain your equations with our definition of multiplication.

Button Strip C has 75 buttons, and Button Strip D has 120 buttons.

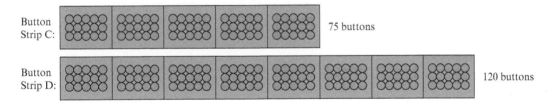

3. How many groups of Button Strip C does it take to make the exact same number of buttons as Button Strip D?

    How many groups of Button Strip D does it take to make the exact same number of buttons as Button Strip C?

4. Use your answers to part 3 to write multiplication equations that relate 75 and 120. Explain your equations with our definition of multiplication.

SECTION 5.1 CLASS ACTIVITY 5-C 🍎

Extending Multiplication to Fractions, Part II

CCSS CCSS SMP6, 5.NF.4a, 5.NF.6

In this Class Activity, use our definitions of multiplication and of fractions throughout. If you want to so solve the problems, pretend you don't yet know how to compute products with fractions, and reason about your math drawings instead.

Recall how our definition of multiplication works with fractions. For example:

$$\frac{3}{4} \quad \bullet \quad 8 \quad = \quad ?$$

of a group   units   units

$\frac{3}{4}$ *of a* group of 8 units are how many units?

$\frac{3}{4}$ *of* 8 units are how many units?

1. Write a simple word problem and make a math drawing for the example above.

2. If 1 serving of juice has 12 grams of sugar, then how many grams of sugar are in $\frac{1}{4}$ of a serving? Make a math drawing and write and annotate an equation for this problem.

3. Interpret the equation below using our definition of multiplication. Then write a word problem and make and explain a math drawing for it.

$$\frac{1}{3} \cdot 15 = ?$$

4. If 1 serving of juice is $\frac{1}{5}$ liter, then how many liters are in $\frac{1}{3}$ of a serving? Make a math drawing and write and annotate an equation for this problem.

5. Interpret the equation below using our definition of multiplication. Then write a word problem and make and explain a math drawing for it.

$$\frac{2}{5} \cdot \frac{1}{3} = ?$$

6. If 1 serving of frozen slurpy is $\frac{4}{5}$ of a liter, then how many liters are in $\frac{2}{3}$ of a serving? Make a math drawing and write and annotate an equation for this problem.

7. Interpret the equation below using our definition of multiplication. Then write a word problem and make and explain a math drawing for it.

$$\frac{4}{5} \cdot \frac{2}{3} = ?$$

8. The math drawing below shows a road made of 7 equal sections. If a 5-section portion of the road is $\frac{3}{4}$ of a mile long, then how long is the entire road? Write and annotate a multiplication equation for this problem.

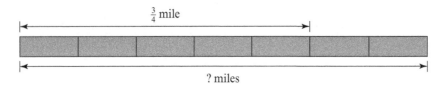

9. Interpret the equation below using our definition of multiplication. Then use a math drawing to explain why the product is 1.

$$\frac{5}{3} \cdot \frac{3}{5} = ?$$

SECTION 5.1 **CLASS ACTIVITY 5-D** 🍎

Explaining Why the Procedure for Multiplying Fractions Is Valid

CCSS CCSS SMP3, 5.NF.4a

1. For each of the rectangles below, fill in the blanks so you can use multiplication to describe the fraction of the area of the rectangle that is shaded. You may assume that all lengths that appear to be the same really are the same.

(a)

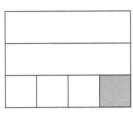

_____ of _____ is shaded.

_____ × _____ is shaded.

(b)

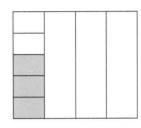

_____ of _____ is shaded.

_____ × _____ is shaded.

(c)

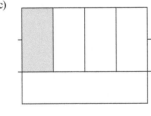

_____ of _____ is shaded.

_____ × _____ is shaded.

(d)

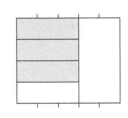

_____ of _____ is shaded.

_____ × _____ is shaded.

2. Interpret $\frac{1}{4} \cdot \frac{1}{3}$ using our definition of multiplication. Then use the math drawings below to explain why

$$\frac{1}{4} \cdot \frac{1}{3} = \frac{1}{4 \cdot 3}$$

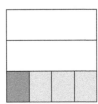
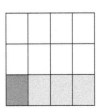

3. Interpret $\frac{2}{3} \cdot \frac{5}{8}$ using our definition of multiplication. Then use the math drawings below to explain why

$$\frac{2}{3} \cdot \frac{5}{8} = \frac{2 \cdot 5}{3 \cdot 8}$$

In particular, explain why we multiply the denominators and why we multiply the numerators.

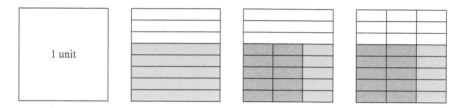

4. Use our definition of multiplication and math drawings to explain why

$$\frac{2}{5} \cdot \frac{3}{7} = \frac{2 \cdot 3}{5 \cdot 7}$$

In particular, explain why we multiply the denominators and why we multiply the numerators.

SECTION 5.1   CLASS ACTIVITY 5-E 🍎

When Do We Multiply Fractions?

CCSS   CCSS SMP2, SMP4

As a teacher, you will probably write word problems for your students. This activity will help you see how slight changes in the wording of a problem can produce big changes in meaning and in the operation that is used to solve the problem.

1. *A mulch pile problem:* Originally, there was $\frac{3}{4}$ of a cubic yard of mulch in a mulch pile. Then $\frac{1}{3}$ of the mulch in the mulch pile was removed. Now how much mulch is left in the mulch pile?

    a. Is the mulch pile problem a problem for $\frac{1}{3} \cdot \frac{3}{4}$, is it a problem for $\frac{3}{4} - \frac{1}{3}$, or is it a problem for neither of these? Explain.

    b. Write a new mulch pile problem for $\frac{1}{3} \cdot \frac{3}{4}$ and write a new mulch pile problem for $\frac{3}{4} - \frac{1}{3}$. Make clear which is which.

2. Which of the following problems are word problems for $\frac{2}{3} \cdot \frac{1}{4}$, and which are not? Why?

    a. Joe is making $\frac{2}{3}$ of a recipe. The full recipe calls for $\frac{1}{4}$ cup of water. How much water should Joe use?

    b. $\frac{1}{4}$ of the students in Mrs. Watson's class are doing a dinosaur project. $\frac{2}{3}$ of the children doing the dinosaur project have completed it. How many children have completed a dinosaur project?

    c. $\frac{1}{4}$ of the students in Mrs. Watson's class are doing a dinosaur project. $\frac{2}{3}$ of the children doing the dinosaur project have completed it. What fraction of the students in Mrs. Watson's class have completed a dinosaur project?

    d. There is $\frac{1}{4}$ of a cake left in Mrs. Watson's class. $\frac{2}{3}$ of the class would like to have some cake. What fraction of the cake does each student who wants cake get?

    e. Carla is making snack bags that each contain $\frac{1}{4}$ package of jelly worms. $\frac{2}{3}$ of her snack bags have been purchased. What fraction of her jelly worms have been purchased?

SECTION 5.1   CLASS ACTIVITY 5-F

What Fraction Is Shaded?

CCSS   CCSS SMP1, SMP7

1. For each of the figures below, write an expression that uses both multiplication and addition (or subtraction) to describe the total fraction of the area of the figure that is shaded. (For example, $\frac{5}{7} \cdot \frac{2}{9} + \frac{1}{3}$ is an expression that uses both multiplication and addition). Explain your reasoning. Then compute that fraction (in simplest form). In each figure, you may assume that lengths appearing to be equal really are equal.

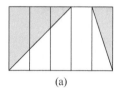

(a)

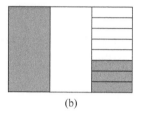

(b)

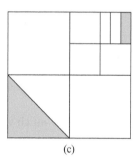

(c)

2. Draw a figure in which you shade $\frac{1}{3} \cdot \frac{2}{7} + \frac{1}{2} \cdot \frac{3}{7}$ of the figure.

SECTION 5.2 **CLASS ACTIVITY 5-G**

Decimal Multiplication and Estimation

CCSS CCSS SMP2

1. Ben wants to multiply 3.46×1.8. He first multiplies the numbers by ignoring the decimal points:

$$
\begin{array}{r}
3.46 \\
\times\ 1.8 \\
\hline
6228
\end{array}
$$

Ben knows that he needs to figure out where to put the decimal point in his answer, but he can't remember the rule about where to put the decimal point. Explain how Ben can reason about the sizes of the numbers to determine where to put the decimal point in his answer.

2. What if Ben's original problem in part 1 was 3.46×0.18? How can he then reason about the sizes of the numbers to determine where to put the decimal point in his answer?

3. Using a calculator, Lameisha finds that

$$1.5 \times 1.2 = 1.8$$

She wants to know why the rule about adding the number of places behind the decimal point doesn't work in this case. Why aren't there 2 digits to the right of the decimal point in the answer? Is Lameisha right that the rule about adding the number of places behind the decimal points doesn't work in this case? Explain.

SECTION 5.2   CLASS ACTIVITY 5-H 🎋

Explaining Why We Place the Decimal Point Where We Do When We Multiply Decimals

CCSS   CCSS SMP7, 5.NBT.7

1. As indicated below, to get from 1.36 to 136, we multiply by 10×10. To get from 2.7 to 27, we multiply by 10. In other words,

$$136 = 10 \times 10 \times 1.36 \quad \text{and} \quad 27 = 10 \times 2.7$$

$$
\begin{array}{r}
1.36 \\
\times\ 2.7
\end{array}
\xrightarrow[\ \times 10\ ]{\ \times 10\ \times 10\ }
\begin{array}{r}
136 \\
\times\ \ 27 \\
\hline
952 \\
2720 \\
\hline
3672
\end{array}
$$

Therefore,

$$
\begin{array}{r}
1.36 \\
\times\ 2.7 \\
\hline
952 \\
2720 \\
\hline
\end{array}
\xleftarrow[\ \ ?\ \ ]{\ \ ?\ \ }
\begin{array}{r}
136 \\
\times\ \ 27 \\
\hline
952 \\
2720 \\
\hline
3672
\end{array}
$$

Explain what we should do now to

$$136 \times 27 = 3672$$

to get back to

$$1.36 \times 2.7$$

Use your answer to explain the placement of the decimal point in 1.36×2.7.

2. More generally, explain why the following is valid: If you multiply a number that has 3 digits to the right of its decimal point by a number that has 4 digits to the right of its decimal point, you should place the decimal point $3 + 4 = 7$ places from the end of the product calculated without the decimal points.

SECTION 5.2   CLASS ACTIVITY 5-I

Decimal Multiplication Word Problems

1. Write a word problem for $2.7 \times 1.35 = ?$ and use our definition of multiplication to explain why it is a problem for that equation.

2. For each of the following problems, determine if it is a decimal multiplication problem. If so, use our definition of multiplication to explain why and write the corresponding equation. If not, write a new decimal multiplication problem that uses a similar context, use our definition of multiplication to explain why it is a multiplication problem, and write the corresponding equation.

   a. If 1 cubic meter of gravel weighs 1.68 tonnes, then how much do 3.8 cubic meters of gravel weigh?

   b. If you can exchange 0.91 euros for $1, then what is the value of 25.63 euros in dollars?

   c. If you can exchange $1 for 23.67 Mexican pesos, then what is the value of $35.45 in Mexican pesos?

   d. Given that 2.2 pounds are equivalent to 1 kilogram, how many pounds are 5.5 kilograms?

   e. If 3.5 liters of a liquid weigh 2.5 kilograms, then how much does 1 liter of the liquid weigh?

SECTION 5.2   **CLASS ACTIVITY 5-J**

Decimal Multiplication and Areas of Rectangles

CCSS   CCSS SMP2

1. Find the area of the 2.3-unit-by-1.8-unit rectangle *without* multiplying. Explain. Then calculate 2.3 × 1.8 and verify that you get the same answer.

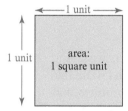

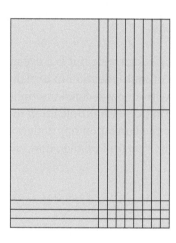

2. Find the area of the 2.4-unit-by-3.6-unit rectangle *without* multiplying. Explain. Then verify that you get the same answer when you multiply 2.4 × 3.6.

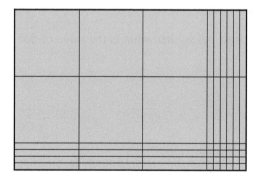

3. Discuss the following questions: How is a 2.4-unit-by-3.6-unit rectangle related to an array or rectangle for 24 × 36? How is 2.4 × 3.6 related to 24 × 36?

   How are decimal, whole number, and mixed number multiplication similar?

SECTION 5.3   **CLASS ACTIVITY 5-K**

Using the Distributive Property to Explain Multiplication with Negative Numbers (and 0)

CCSS   CCSS SMP3, 7.NS.2a,c

1. Write a word problem for $3 \cdot -5$. Solve the word problem, thereby explaining why $3 \cdot -5$ is negative. You might interpret negative numbers as amounts owed or as negatively charged particles.

2. Explain why the following make sense:

$$0 \cdot (\text{any number}) = 0$$
$$(\text{any number}) \cdot 0 = 0$$

3. Assume that you don't yet know what $(-3) \cdot 5$ is, but you do know that $3 \cdot 5 = 15$. Use the distributive property to show that the expression

$$(-3) \cdot 5 + 3 \cdot 5$$

is equal to 0. Then use that result to determine what $(-3) \cdot 5$ must be equal to.

4. Assume that you don't yet know what $(-3) \cdot (-5)$ is, but you do know that $(-3) \cdot 5 = -15$ from part 3. Use the distributive property to show that the expression

$$(-3) \cdot (-5) + (-3) \cdot 5$$

is equal to 0. Then use that result to determine what $(-3) \cdot (-5)$ must be equal to.

SECTION 5.4 **CLASS ACTIVITY 5-L**

Multiplying Powers of 2

CCSS CCSS SMP3, SMP8, 8.EE.1

1. Use the meaning of powers of 2 to explain how to write each of the expressions below as a single power of 2. For example:

$$2^3 \times 2^2 = (2 \times 2 \times 2) \times (2 \times 2) = 2^5$$

   a. $2^4 \times 2^3$

   b. $2^5 \times 2^6$

   c. $2^6 \times 2^3$

2. Looking at your work in part 1, how is the exponent of the product related to the exponents of the factors? (For example, in the example in part 1, how are the exponents 3 and 2 related to the exponent 5?)

3. Explain why it is always true that $2^A \times 2^B = 2^{A+B}$ whever A and B are counting numbers.

Assume that the equation $2^A \times 2^B = 2^{A+B}$ is true not just when A and B are counting numbers, but for all numbers.

4. Based on the assumption above, what must the following be equal to?

$2^5 \times 2^0$

$2^7 \times 2^0$

$2^A \times 2^0$

5. Based on your answer in part 4, what must 2^0 be equal to?

6. Based on your answer in part 5, what must the following be equal to?

$2^3 \times 2^{-3}$

$2^5 \times 2^{-5}$

$2^A \times 2^{-A}$

7. Based on your answer in part 6, what must the following be equal to?

2^{-3}

2^{-5}

8. Based on your answers in parts 6 and 7, explain why the following equation should be true:

$$2^{-A} = \frac{1}{2^A}$$

9. Based on part 8 and the fact that $A - B = A + (-B)$, how can you express 2^{A-B} in terms of 2^A and 2^B?

SECTION 5.4   **CLASS ACTIVITY 5-M**

Multiplying Powers of 10

CCSS   CCSS SMP3, SMP8, 8.EE.1

1. Use the meaning of powers of 10 to show how to write each of the expressions in (a), (b), and (c) as a single power of 10 (i.e., in the form 10^A for some exponent A). For example, 10^2 means 10×10, and 10^3 means $10 \times 10 \times 10$; therefore,

$$10^2 \times 10^3 = (10 \times 10) \times (10 \times 10 \times 10) = 10^5$$

   a. $10^3 \times 10^4$

   b. $10^2 \times 10^5$

   c. $10^3 \times 10^3$

2. In (a), (b), and (c) in part 1, relate the exponents in the product with the exponent in the answer. (For the example given at the beginning of part 1, relate 2 and 3 to 5.) In each case, how are the three exponents related?

3. Explain why it is always true that $10^A \times 10^B = 10^{A+B}$ when A and B are counting numbers.

4. Assume now that we want the equation $10^A \times 10^B = 10^{A+B}$ to be true not just when A and B are counting numbers, but even when A or B is 0. With this assumption, explain why it makes sense that 10^0 should be equal to 1.

5. Assume now that we want the equation $10^A \times 10^B = 10^{A+B}$ to be true not just when A and B are whole numbers, but for all numbers. Also assume that $10^0 = 1$.

With these assumptions, explain why the following make sense:

a. 10^{-1} should be equal to $\frac{1}{10}$.

b. 10^{-2} should be equal to $\frac{1}{100}$.

c. 10^{-N} should be equal to $\frac{1}{10^N}$.

SECTION 6.1   CLASS ACTIVITY 6-A 🍎

What Does Division Mean?

CCSS CCSS 3.OA.2

1. Write a simple word problem and make a math drawing that you could use to help children understand what $10 \div 2$ means.

2. Reformulate the division problem $10 \div 2 = ?$ as

$$2 \times ? = 10 \quad \text{or as} \quad ? \times 2 = 10$$

whichever fits with your word problem in part 1 and with the way we have described multiplication.

3. Now write another simple division word problem for $10 \div 2$, one that fits with the *other* multiplication equation identified in part 2. Make a math drawing that fits with your new problem.

SECTION 6.1   CLASS ACTIVITY 6-B 🍎

Division Word Problems

CCSS CCSS SMP4, 3.OA.2

For each problem, make a math drawing, determine if it is a how-many-groups or a how-many-units-in-1-group division problem, and write annotated division and multiplication equations. Discuss how the annotated division equations are similar for the how-many-groups problems and how they are similar for the how-many-units-in-1-group problems.

How-many-groups division	How-many-units-in-one-group division
There are 15 stickers to put in bags. 5 stickers go in each bag. How many bags do we need?	There are 15 stickers to distribute equally among 5 bags. How many stickers go in each bag?
1 bag ⟷ 1 group = 5 stickers	1 bag ⟷ 1 group = ? stickers
? • **5** = **15**	**5** • **?** = **15**
groups stickers stickers	groups stickers stickers
How many groups of 5 stickers are 15 stickers?	5 groups of how many stickers are 15 stickers?

1. Bill has a muffin recipe that calls for 2 cups of flour. How many batches of muffins can Bill make if he has 8 cups of flour available? (Assume that Bill has sufficient amounts of all the other ingredients.)

2. One foot is 12 inches. If a piece of rope is 96 inches long, then how long is it in feet?

3. Francine has 32 yards of rope that she wants to cut into 8 equal pieces. How long will each piece be?

4. If 1 gallon of water weighs 8 pounds, how many gallons will there be in 400 pounds of water?

5. If you drive 220 miles at a constant speed and it takes you 4 hours, then how fast did you drive?

6. If 6 limes cost $3, then how many limes should you be able to buy for $1?

7. If 4 avocados cost $5, then how much should 1 avocado cost?

8. Write a situation for $4 \cdot 8 = 32$. Then write two related division word problems, one for each of the two types of division.

Why Can't We Divide by Zero?

CCSS   CCSS SMP2, 7.NS.2b

1. Use the fact that every division problem can be rewritten as a multiplication problem with a factor unknown to explain why $2 \div 0$ is not defined.

2. Write word problems for the two interpretations of $2 \div 0 = ?$ Use your problems to explain why $2 \div 0$ is not defined. Link your word problems to your multiplication equations from part 1.

3. Write word problems for the two interpretations of $0 \div 2 = ?$ Use your problems to explain why $0 \div 2$ *is* defined. Explain the difference between $2 \div 0$ and $0 \div 2$.

4. Explain why $0 \div 0$ is undefined by viewing division in terms of multiplication. (We can also say that $0 \div 0$ is "indeterminate.") Can you give the same explanation as for why $2 \div 0$ is not defined? If not, how are the explanations different?

SECTION 6.2 | **CLASS ACTIVITY 6-D**

Relating Whole Number Division and Fractions

CCSS CCSS SMP6, SMP8, 5.NF.3

1. There are 3 identical pizzas that will be divided equally among 4 people. How much pizza will each person get? Use a math drawing to solve the problem and explain your answer. What does your answer tell you about $3 \div 4$?

2. There are 4 liters of juice to be shared equally among 5 people. How much juice does each person get? Explain why this is a problem for $4 \div 5 = ?$ and use a math drawing to explain why the solution is $\frac{4}{5}$ of a liter. Therefore, how are $4 \div 5$ and $\frac{4}{5}$ related?

3. In general, why is it true that $A \div B = \frac{A}{B}$ whenever A and B are counting numbers?

SECTION 6.2   **CLASS ACTIVITY 6-E**

Using Measurement to Relate Whole Number Division and Fractions

CCSS   CCSS SMP8, 5.NF.3

1. Measurement problem: How many of the first strip does it take to make the second strip?

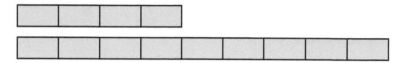

   a. Explain how to interpret the measurement problem as asking $? \cdot 4 = 9$ and $9 \div 4 = ?$

   b. Solve the measurement problem by reasoning about the strips with our definition of fraction.

   c. Repeat parts (a) and (b) but now with this measurement problem: How many of the second strip does it take to make the first strip? (You will need to modify the equations in part (a).)

2. Use how-many-groups division and the idea of measuring one strip by another strip to explain why $3 \div 7 = \frac{3}{7}$ and why $7 \div 3 = \frac{7}{3}$.

SECTION 6.2 **CLASS ACTIVITY 6-F** 🍎

What to Do with the Remainder?

CCSS CCSS SMP4, SMP6, 4.OA.3

1. Consider these two word problems for $14 \div 3$:

   A baking problem: A batch of cookies requires 3 cups of flour. How many batches of cookies can you make if you have 14 cups of flour (and all the other ingredients you need)?

   A brownie problem: You have 14 brownies which you will divide equally among 3 bags. How many brownies should you put in each bag?

   a. In the table below, write your interpretation of the whole number quotient 4, remainder 2, and the mixed number quotient $4\frac{2}{3}$ for both problems. In each case, what does the 4 stand for? What does the 2 stand for? What does the $\frac{2}{3}$ stand for? Could the $\frac{2}{3}$ stand for something else (not connected to 4)?

	Baking problem: 1 batch ⟶ 3 cups ? batches ⟶ 14 cups	Brownie problem: 3 bags ⟶ 14 brownies 1 bag ⟶ ? brownies
4, R 2		
$4\frac{2}{3}$		

   b. What is different about how the whole number quotient 4 and remainder 2 are interpreted in the two problems?

   c. Discuss how the remainder 2 and the $\frac{2}{3}$ are related for each problem.

2. Write a word problem for which you would calculate 14 ÷ 3 to solve the problem, but which has the answer 5.

3. Write a word problem for which you would calculate 14 ÷ 3 to solve the problem, but which has answer 2, the remainder.

4. *A calendar problem:* What day of the week will it be 31 days from today? Explain how 31 ÷ 7 is relevant to solving the calendar problem.

5. Consider these three problems about distance, speed, and time:
    i. How long will it take you to drive 180 miles if you drive at the constant speed of 55 mph?

    ii. How long will it take you to drive 195 miles if you drive at the constant speed of 60 mph?

    iii. How long will it take you to drive 105 miles if you drive at the constant speed of 30 mph?

    a. Write numerical division problems to solve problems (i), (ii), and (iii) and give the mixed number answers and whole-number-with-remainder answers to these numerical problems.

    b. Interpret the meaning of the mixed number answers and whole-number-with-remainder answers you gave in part (a) in terms of the original word problems.

    c. For problem (i) Josh says: "The 3, remainder 15, answer tells you that it will take 3 full hours, and the remainder 15 tells you it will take another 15 minutes." Explain why Josh's comment is approximately correct, but not completely correct.

Discuss Division Reasoning

CCSS CCSS SMP3, 4.NBT.6, 5.NBT.6

1. There are 260 pencils to be put in packages of 12. How many packages of pencils can be made, and how many pencils will be left over?

Antrice's solution:

10 packages will use up 120 pencils. After another 10 packages, 240 pencils will be used up. After 1 more package, 252 pencils are used. Then there are only 8 pencils left, and that's not enough for another package. So the answer is 21 packages of pencils with 8 pencils left over.

Explain why the equations below correspond to Antrice's work, and explain why the last equation shows that 260 ÷ 12 has whole number quotient 21, remainder 8.

$$(10 \cdot 12) + (10 \cdot 12) + (1 \cdot 12) + 8 = 260$$
$$(10 + 10 + 1) \cdot 12 + 8 = 260$$
$$21 \cdot 12 + 8 = 260$$

2. Ashley's work on the division problem 258 ÷ 6 is shown below. Explain what Ashley did and why her strategy makes sense. Then write equations that correspond to Ashley's work and demonstrate that 258 ÷ 6 = 43.

$$258 \div 6 = ?$$

10 ⟶ 60	240 ⟵ 40	
20 ⟶ 120	+12 ⟵ 2	
40 ⟶ 240	252 ⟵ 42	
	+ 6	
	258 ⟵ 43	

3. Zane's work on the division problem 245 ÷ 15 is shown below. Explain why Zane's strategy makes sense. Then write equations that correspond to Zane's work and demonstrate that 245 ÷ 15 has whole number quotient 16, remainder 5.

$$
\begin{array}{r}
15 \\
\times\ 2 \\
\hline
30 \\
\times\ 2 \\
\hline
60 \\
\times\ 4 \\
\hline
240 \\
5\ \text{left}
\end{array}
$$

$2 \times 2 \times 4 = 16\,\mathrm{R}5$

4. Pretend that you don't have a calculator and have forgotten how to do longhand division. Explain how you can calculate 5170 ÷ 6.

SECTION 6.3 CLASS ACTIVITY 6-H

Why the Scaffold Method of Division Works

CCSS CCSS SMP3, 4.NBT.6

1. Interpret each of the steps in the next scaffold in terms of the following word problem:

    You have 3475 marbles, and you want to put these marbles into bags with 8 marbles in each bag. How many bags of marbles can you make, and how many marbles will be left over?

$$
\begin{array}{r}
4 \\
30 \\
400 \\
8)\overline{3475} \\
-\,3200 \\
\hline
275 \\
-\,240 \\
\hline
35 \\
-\,32 \\
\hline
3
\end{array}
$$

    Then relate these equations to the scaffold and the division problem:

$$3475 - 400 \cdot 8 - 30 \cdot 8 - 4 \cdot 8 = 3$$
$$3475 - (400 + 30 + 4) \cdot 8 = 3$$
$$3475 - 434 \cdot 8 = 3$$

2. Use the scaffold method to calculate $8321 \div 6$. (You may use the method flexibly or in standard algorithm form.) Interpret each step in your scaffold in terms of the following word problem:

    You have 8321 pickles, and you want to put these pickles in packages with 6 pickles in each package. How many packages can you make, and how many pickles will be left over?

    Then write equations like those in part 1 and relate them to the scaffold and the division problem.

SECTION 6.3   CLASS ACTIVITY 6-I

How Can We Divide Base-Ten Bundles?

CCSS SMP2, SMP5, 4.NBT.6

Materials Each group will need about 260 toothpicks (or other small objects that can be bundled) and 20 rubber bands.

1. **a.** Organize 97 toothpicks into base-ten bundles: 9 bundles of ten and 7 individual toothpicks.

    b. Divide the 97 toothpicks equally among 4 groups and record your method (step 1, . . . step 2, . . . , etc.).

    c. If it fits, write some notation that captures or corresponds to the steps you took in part (b).

    d. Can you think of a different set of steps for dividing the 97 toothpicks equally among the 4 groups?

2. **a.** Organize 256 toothpicks into base-ten bundles: 2 bundles of a hundred, *each of which is 10 bundles of ten;* 5 bundles of ten; and 6 individual toothpicks.

    b. Divide the 256 toothpicks equally among 3 groups and record your method (step 1, . . . step 2, . . . , etc.).

    c. If it fits, write some notation that captures or corresponds to the steps you took in part (b).

    d. Can you think of a different set of steps for dividing the 256 toothpicks equally among the 3 groups?

SECTION 6.3   CLASS ACTIVITY 6-J 🍎

Interpreting the Standard Division Algorithm as Dividing Base-Ten Bundles

CCSS   CCSS SMP3, SMP5, 4.NBT.6

For each part in this activity, use the *common method* for implementing the standard algorithm. Explain how to interpret each step in terms of dividing bundled toothpicks equally among a number of groups. View the toothpicks as ones, bundles of tens, bundles of hundreds, *which are 10 bundles of ten*, and bundles of 1000s, *which are 10 bundles of one hundred*.

Pay special attention to these points:

- Do not say "goes into." Instead of "4 goes into 9," you can say, "There are 4 groups, and we have 9 bundles of ten."
- Interpret the "bringing down" steps as unbundling and combining with the bundles at the next lower place.

1. $4\overline{)93}$

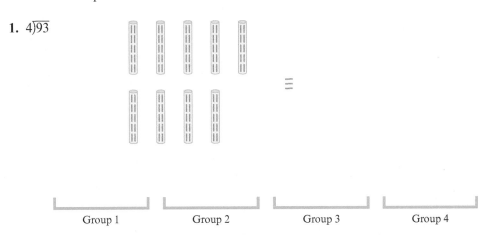

Group 1   Group 2   Group 3   Group 4

2. $3\overline{)143}$

3. $3\overline{)1372}$

4. $6\overline{)1230}$

SECTION 6.3   CLASS ACTIVITY 6-K 🏺

Interpreting the Standard Division Algorithm in Terms of Money

CCSS   CCSS SMP3, 5.NBT.7

1. Use the standard division algorithm to determine the decimal answer to 2674 ÷ 3 to the hundredths place. Interpret each step in your calculation in terms of dividing $2674 equally among 3 people by imagining that you distribute the money in stages: First distribute hundreds, then tens, then ones, then dimes (tenths), then pennies (hundredths).

2. Use the standard division algorithm to determine the decimal answer to 125 ÷ 6 to the hundredths place. Interpret each step in your calculation in terms of dividing $125 equally among 6 people in stages, as in part 1.

SECTION 6.3 | CLASS ACTIVITY 6-L

Critique Reasoning about Decimal Answers to Division Problems

CCSS CCSS SMP3

1. Here is how Ben answered some division problems:

$$251 \div 6 = 41.5$$
$$269 \div 7 = 38.3$$
$$951 \div 21 = 45.6$$

Is Ben right? How might Ben be thinking?

2. If you know that the answer to a whole number division problem is 4, remainder 1, can you tell what the decimal answer to the division problem is without any additional information? If not, what other information would you need to determine the decimal answer?

SECTION 6.4   CLASS ACTIVITY 6-M 🍎

How-Many-Groups Fraction Division Problems

CCSS   CCSS SMP3, SMP6, 5.NF.7, 6.NS.1

Example for $1\frac{1}{2} \div \frac{1}{4} = ?$

If 1 serving of steel cut oats is $\frac{1}{4}$ of a cup and you have $1\frac{1}{2}$ cups left, then how many servings of steel cut oats do you have left?

1 serving ⟷ 1 group = $\frac{1}{4}$ of a cup

$$\underset{\text{groups}}{\textbf{?}} \quad \bullet \quad \underset{\text{of a cup}}{\frac{\mathbf{1}}{\mathbf{4}}} \quad = \quad \underset{\text{cups}}{\mathbf{1\frac{1}{2}}}$$

How many groups of $\frac{1}{4}$ make $1\frac{1}{2}$?

You may use equivalent fractions wherever they help you.

1. Explain how to use a math drawing to solve the problem in the example.

2. Write a simple how-many-groups word problem for $3 \div \frac{3}{4} = ?$ and explain how to solve the problem with the aid of a math drawing.

3. Tonya and Chrissy are trying to understand $1 \div \frac{2}{3} = ?$ by using the following problem:

One serving of rice is $\frac{2}{3}$ of a cup. I ate 1 cup of rice. How many servings of rice did I eat?

To solve the problem, Tonya and Chrissy draw a square partitioned into three equal pieces, and they shade two of those pieces.

Tonya says, "There is one $\frac{2}{3}$-cup serving of rice in 1 cup, and there is $\frac{1}{3}$ cup of rice left over, so the answer should be $1\frac{1}{3}$."

Chrissy says, "The part left over is $\frac{1}{3}$ cup of rice, but the answer is supposed to be $\frac{3}{2} = 1\frac{1}{2}$. Did we do something wrong?"

Help Tonya and Chrissy.

4. Write a how-many-groups word problem for $1\frac{1}{2} \div \frac{1}{3} = ?$ and solve your problem with the aid of a math drawing, a table, or a double number line. Explain your reasoning.

5. Write a how-many-groups word problem for $\frac{1}{3} \div \frac{3}{4} = ?$ and solve your problem with the aid of a math drawing, a table, or a double number line. Explain your reasoning.

Reasoning about Equivalent Problems to Divide Fractions That Have Common Denominators

CCSS CCSS SMP4, SMP8, 6.NS.1

Measurement problem: How many of the first strip does it take to make the second strip?

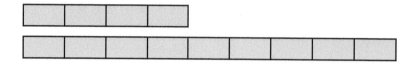

1. Explain how to interpret *each* of the following equations as modeling the measurement problem. To do so, you will need to suitably interpret (and reinterpret) the size of each part in the strips above.

    a. $\frac{9}{3} \div \frac{4}{3} = ?$ In this case, what is the size of each part?

    b. $\frac{9}{5} \div \frac{4}{5} = ?$ In this case, what is the size of each part?

    c. $\frac{9}{8} \div \frac{4}{8} = ?$ In this case, what is the size of each part?

    d. $9 \div 4 = ?$ In this case, what is the size of each part?

2. Solve the measurement problem by reasoning about the strips. In solving the problem, does it matter what we say the size of each part is (as long as all the parts are the same size)? What does that tell you about the solutions to the division problems in part 1?

3. Write one measurement problem about strips so that *each* of the following equations models the problem. How can four different equations model the same problem?

$$\frac{3}{7} \div \frac{5}{7} = ? \qquad \frac{3}{8} \div \frac{5}{8} = ? \qquad \frac{3}{2} \div \frac{5}{2} = ? \qquad 3 \div 5 = ?$$

4. Solve your measurement problem in part 3 by reasoning about the strips. What can you conclude about the solutions to the division problems in part 3?

5. Reflecting on parts 1–4, in general, how are the division problems

$$\frac{A}{C} \div \frac{B}{C} = ? \quad \text{and} \quad A \div B = ?$$

related? Therefore, what is a method for dividing fractions that have a common denominator?

SECTION 6.4 **CLASS ACTIVITY 6-O**

Dividing Fractions by Dividing the Numerators and Dividing the Denominators

CCSS CCSS SMP3, 6.NS.1

1. Consider the two division problems

$$\frac{6}{5} \div \frac{2}{5} = ? \quad \text{and} \quad 6 \div 2 = ?$$

Explain in two ways why these division problems must have the same solution:

- By interpreting the small rectangles below in two ways.

How many ⬚⬚ are in ⬚⬚⬚⬚⬚⬚ ?

- By rewriting the division problems as multiplication problems with unknown factors.

2. View

$$\frac{6}{20} \div \frac{3}{4} = \frac{?}{?} \quad \text{as} \quad \frac{?}{?} \cdot \frac{3}{4} = \frac{6}{20}$$

and explain how to deduce that

$$\frac{6}{20} \div \frac{3}{4} = \frac{6 \div 3}{20 \div 4} = \frac{2}{5}$$

3. Give another example where you can divide fractions by dividing the numerators and dividing the denominators. Use the reasoning of part 2 to explain why this method works.

4. Explain how to use equivalent fractions so that you can apply the method of parts 2 and 3 to other cases, such as

$$\frac{5}{7} \div \frac{3}{4}$$

SECTION 6.5 **CLASS ACTIVITY 6-P**

Why Is Dividing by 𝓐 Equivalent to Multiplying by $\frac{1}{A}$?

CCSS CCSS SMP2, SMP4, SMP8, 5.NF.4a

1. A company puts beads on strips of different lengths. Each section has the same number of beads. Write and explain several multiplication and division equations to show how the numbers of beads on the strip and the part are related. Include an equation that involves a fraction.

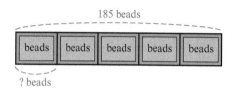

185 beads

| beads | beads | beads | beads | beads |

? beads

2. Based on part 1, how are $185 \div 5$ and $\frac{1}{5} \cdot 185$ related? Explain!

3. A company sells ropes in different lengths. All the sections of rope weigh the same. Write and explain several multiplication and division equations to show how the weights of the rope and a section are related. Include an equation that involves a fraction.

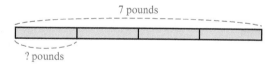

7 pounds

? pounds

4. Based on part 3, how are $7 \div 4$ and $\frac{1}{4} \cdot 7$ related? Explain!

5. Explain why $17 \div 3 = \frac{1}{3} \cdot 17$ with the aid of a situation and a math drawing.

6. By reflecting on the previous parts, explain more generally why dividing a number by a natural number A is equivalent to multiplying the number by $\frac{1}{A}$.

SECTION 6.5   CLASS ACTIVITY 6-Q

Why Is Dividing by $\frac{A}{B}$ Equivalent to Multiplying by $\frac{B}{A}$?

CCSS   CCSS SMP4, 5.NF.7, 6.NS.1

1. *Road problem:* So far, $\frac{1}{3}$ of a road has been paved and this paved portion is $5\frac{1}{2}$ miles long. How long is the entire road?

   a. Make a math drawing for the *Road problem* and solve the problem.

   b. Explain how to interpret the equation $5\frac{1}{2} \div \frac{1}{3} = ?$ as modeling the *Road problem* from the how-many-units-in-1-group perspective.

   c. Explain how to interpret the equation $3 \cdot 5\frac{1}{2} = ?$ as modeling the *Road problem* based on your math drawing.

2. *Sidewalk problem:* So far, $\frac{3}{8}$ of a sidewalk has been built and this portion is $1\frac{1}{2}$ miles long. How long will the entire sidewalk be?

   a. Solve the *Sidewalk problem* by reasoning about the math drawing.

   b. Explain how to interpret the equation $1\frac{1}{2} \div \frac{3}{8} = ?$ as modeling the *Sidewalk problem* from the how-many-units-in-1-group perspective.

   c. Explain how to interpret the equation $\frac{8}{3} \cdot 1\frac{1}{2} = ?$ as modeling the *Sidewalk problem* based on the math drawing.

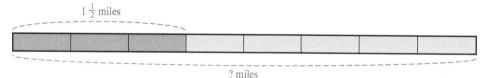

3. a. Write a how-many-units-in-1-group problem for $8 \div \frac{2}{5} = ?$ and explain why it is that type of problem.

   b. Explain how to reason about a math drawing to solve your problem from part (a).

   c. Use your math drawing to explain how to interpret the equation $\frac{5}{2} \cdot 8 = ?$ as modeling the problem you wrote in part (a).

4. Reflecting on parts 1, 2, and 3, when we think of division from a how-many-units-in-1-group perspective, why is dividing by a fraction $\frac{A}{B}$ equivalent to multiplying by the fraction $\frac{B}{A}$?

SECTION 6.5 CLASS ACTIVITY 6-R 🍎

How-Many-Units-in-1-Group Fraction Division Problems

CCSS CCSS SMP3, SMP6, 5.NF.7, 6.NS.1

Example for $15 \div \frac{3}{5} = ?$

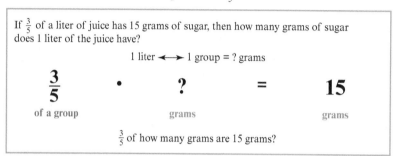

If $\frac{3}{5}$ of a liter of juice has 15 grams of sugar, then how many grams of sugar does 1 liter of the juice have?

1 liter ←——→ 1 group = ? grams

$$\frac{3}{5} \quad \bullet \quad ? \quad = \quad 15$$

of a group   grams   grams

$\frac{3}{5}$ of how many grams are 15 grams?

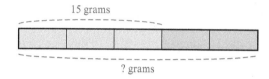

15 grams

? grams

1. Solve the problem in the example. Then explain how to see the solution as multiplying 15 by the reciprocal of $\frac{3}{5}$, or in other words as $\frac{5}{3} \cdot 15$.

2. Write a how-many-units-in-1-group problem for $3 \div \frac{1}{4} = ?$ and explain how to solve the problem with the aid of a math drawing.

   Use your math drawing to explain why $3 \div \frac{1}{4}$ is equivalent to multiplying 3 by the reciprocal of $\frac{1}{4}$, in other words to $\frac{4}{1} \cdot 3$.

3. Write a how-many-units-in-1-group problem for $6 \div \frac{3}{4} = ?$ and explain how to solve the problem with the aid of a math drawing.

   Use your math drawing to explain why $6 \div \frac{3}{4}$ is equivalent to $\frac{4}{3} \cdot 6$.

4. Write a how-many-units-in-1-group problem for $\frac{1}{4} \div \frac{2}{3} = ?$ and explain how to solve the problem with the aid of a math drawing. You may use equivalent fractions.

   Use your math drawing to explain why $\frac{1}{4} \div \frac{2}{3}$ is equivalent to $\frac{3}{2} \cdot \frac{1}{4}$.

5. Write a how-many-units-in-1-group problem for $\frac{3}{5} \div \frac{2}{3} = ?$ and explain how to solve the problem with the aid of a math drawing. You may use equivalent fractions.

   Use your math drawing to explain why $\frac{3}{5} \div \frac{2}{3}$ is equivalent to $\frac{3}{2} \cdot \frac{3}{5}$.

SECTION 6.5   **CLASS ACTIVITY 6-S**

Are These Division Problems?

CCSS   CCSS SMP4

Which of the following are word problems for the division problem $\frac{3}{4} \div \frac{1}{2}$? For those that are, which interpretation of division is used? For those that are not, determine how to solve the problem if it can be solved.

1. Beth poured $\frac{3}{4}$ cup of cereal in a bowl. The cereal box says that 1 serving is $\frac{1}{2}$ cup. How many servings are in Beth's bowl?

2. Beth poured $\frac{3}{4}$ cup of cereal in a bowl. Then Beth took $\frac{1}{2}$ of that cereal and put it into another bowl. How many cups of cereal are in the second bowl?

3. A crew is building a road. So far, the road is $\frac{3}{4}$ mile long. This is $\frac{1}{2}$ the length that the road will be when it is finished. How many miles long will the finished road be?

4. A crew is building a road. So far, the crew has completed $\frac{3}{4}$ of the road, and this portion is $\frac{1}{2}$ mile long. How long will the finished road be?

5. If $\frac{3}{4}$ cup of flour makes $\frac{1}{2}$ of a batch of cookies, then how many cups of flour are required for a full batch of cookies?

6. If $\frac{1}{2}$ cup of flour makes 1 batch of cookies, then how many batches of cookies can you make with $\frac{3}{4}$ cup of flour?

7. If $\frac{3}{4}$ cup of flour makes 1 batch of cookies, then how much flour is in $\frac{1}{2}$ of a batch of cookies?

Reasoning and Estimation with Decimal Division

CCSS CCSS SMP2

1. Describe a way to calculate $32.5 \div 0.5$ mentally. *Hint:* Think in terms of fractions or in terms of money.

2. Describe ways to calculate $1.2 \div 0.25$ mentally.

3. Describe a way to estimate $7.2 \div 0.333$ mentally.

4. Fran must calculate $2.45 \div 1.5$ longhand, but she can't remember what to do about decimal points. Instead, Fran solves the division problem $245 \div 15$ longhand and gets the answer 16.33. Fran knows that she must shift the decimal point in 16.33 somehow to get the correct answer to $2.45 \div 1.5$. Explain how Fran could use estimation to determine where to put the decimal point.

SECTION 6.6 **CLASS ACTIVITY 6-U** 🍎

How Can We Relate Decimal Division to Whole Number Division?

CCSS CCSS SMP3, 5.NBT.7

1. How are the problems

    "How many groups of $0.25 are in $12.37?" and

    "How many groups of 25 cents are in 1237 cents?"

   related? What can you conclude about how the two division problems

$$0.25\overline{)12.37} \quad \text{and} \quad 25\overline{)1237}$$

   are related?

2. Explain how the figure below can be interpreted as:

$$0.06 \div 0.02 = ?$$

   Explain how the same figure can be interpreted as:

   $0.6 \div 0.2 = ?$ or as $6 \div 2 = ?$ or as $6,000,000 \div 2,000,000 = ?$

   What other division problems can the figure illustrate?

   What's the moral here?

3. Make rough drawings of bundled objects to represent,

    "How many groups of 0.15 are in 1.2?"

   Then describe what other questions your drawing could represent and how it is helpful when calculating $0.15\overline{)1.2}$.

Decimal Division Word Problems

1. Write one how-many-groups word problem and another how-many-units-in-1-group word problem for $23.45 \div 2.7 = ?$.

2. For each of the following, write and explain at least one equation that models the problem, using a question mark for the unknown. If you write a division equation, discuss which kind of division the problem uses (how-many-groups or how-many-units-in-1-group). But watch out: Not every problem is a division problem!

   a. If 1 pound of Spellbinding Nuts cost $5.98, then how many pounds of Spellbinding Nuts should you be able to buy for $19.95?

   b. If 1 liter of Polyjuice Potion weighs 1.45 kilograms, then how many kilograms are 3.7 liters of Polyjuice Potion?

   c. If 1 liter of Polyjuice Potion weighs 1.45 kilograms, then how many liters are 2.63 kilograms of Polyjuice Potion?

   d. If it took 2.6 kilograms of invisible yarn to weave 10.5 square meters of invisible cloth, then how much will the yarn for 1 square meter of invisible cloth weigh?

   e. If 1.5 pounds of Wizard Mix cost $4.76, then how much should 1 pound of Wizard Mix cost?

SECTION 7.1 CLASS ACTIVITY 7-A 🍎

Mixtures: The Same or Different?

CCSS CCSS SMP2, 6.RP.1, 6.RP.3a

Materials Cups and two juices or collections of small square tiles or beads in two colors would be helpful but are optional.

There are two containers, each holding a mixture of 1 cup red punch and 3 cups lemon-lime soda. The first container is left as it is. That is Mixture A. Somebody adds 2 cups red punch and 2 cups lemon-lime soda to the second container. That becomes Mixture B.

Mixture A   Mixture B

1. Do you think Mixture A and Mixture B will taste the same and have the same color? Why or why not? Try to think about these questions in the way that a student who has not yet studied ratios might. What ideas do you think such a student might have?

    If possible, make the mixtures to see if they taste and look the same or not. You can simulate the juice mixtures by mixing cups containing equal numbers of tiles or beads. Use tiles or beads in two different colors.

2. Suppose you mix 4 cups red punch with 12 cups lemon-lime soda. Make a math drawing showing how to *organize* those cups *so that you can tell from the way the cups are organized* that this mixture will have the same flavor and color as Mixture A.

3. Suppose you have 4 identical containers, 1 with red punch and 3 with lemon-lime soda, all poured to the same level. You are about to mix them in a pitcher when you decide you want to make a little bit more. You want this larger amount to have the exact same flavor and color. How can you do that? How will the mixture compare to Mixture A?

4. Find mixtures of red punch and lemon-lime soda that taste and look the same as mixture A. Find other mixtures that taste and look the same as Mixture B. Show these mixtures in the columns of the two tables below.

   Explain why the mixtures in one table will taste and look the same *without using the words "ratio" or "fraction" in your explanation.*

   Mixtures that taste and look the same as Mixture A

Cups of red punch	1				
Cups of lemon-lime	3				
Total number of cups	4				

   Mixtures that taste and look the same as Mixture B

Cups of red punch	3				
Cups of lemon-lime	5				
Total number of cups	8				

5. Explain how to use the tables in part 4 to compare the flavors of mixtures A and B in several ways *without* using the terms "ratio" or "fraction." Which mixture is more red-punchy and which is more lemon-limey? Why is it useful to look for common entries in the two tables? Why is it legitimate to compare different columns from each table?

SECTION 7.1 **CLASS ACTIVITY 7-B**

How Are Ratios Related to Equal Groups?

CCSS CCSS SMP2, SMP7, 6.RP.1

Materials Optionally, each person could use beads in two colors: 10 of one color and 15 of the other.

Let's think about making hot chocolate!

> **Mixture A:** 2 ounces chocolate, 3 ounces milk
>
> **Mixture B:** 8 ounces chocolate, 12 ounces milk

Will these two mixtures taste the same and have the same color?

1. See if you can find a way to organize the ounces of chocolate and milk for Mixture B so you can tell *from the way you organized the ounces* whether the mixture will have the same flavor and color as Mixture A. Make a math drawing to show your organization.

    If you have beads, view each bead as standing for 1 ounce and experiment with ways of organizing the beads.

2. See if you can find mathematical notation that expresses how you organized the ounces of chocolate and milk in part 1.

3. Now repeat parts 1 and 2 but find a different way to organize the ounces.

    If you can't find another way to organize the ounces, then try parts 4, 5, and 6, and return to this part.

4. Suppose you want to make hot chocolate that has the same flavor and color as Mixture A, but with 200 ounces of chocolate. Make a math drawing to show how many ounces of milk you would need.

5. Suppose you want to make hot chocolate that has the same flavor and color as Mixture A, but with 3000 ounces of milk. Make a math drawing to show how many ounces of chocolate you would need.

6. Suppose you want to make just a *tiny bit* more hot chocolate than Mixture A, but still with the same flavor and color as mixture A. How could you do that?

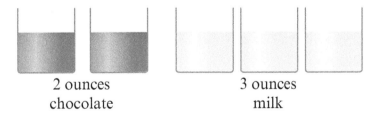

2 ounces
chocolate

3 ounces
milk

SECTION 7.2 | **CLASS ACTIVITY 7-C** 🏺

Using Double Number Lines to Solve Proportion Problems

CCSS CCSS SMP2, SMP3, 6.RP.3

Explain how to solve each of the following problems by reasoning about the quantities. Support your reasoning with double number lines.

1. If 3 yards of rope weigh 2 pounds, then how much do the following lengths of the same kind of rope weigh?

   a. 18 yards   **b.** 16 yards   **c.** 14 yards

2. If 2 meters of wire weigh 24.8 grams, then how much do 15 meters of that same kind of wire weigh? Try to find several ways of reasoning about the quantities to solve this problem.

3. A scooter is going $\frac{3}{4}$ of a mile every 4 minutes. How far does the scooter go in the following amounts of time?

   a. 12 minutes   **b.** 17 minutes

   How long does it take the scooter to go the following distances?

   c. 1 mile   **d.** 2 miles

SECTION 7.2 CLASS ACTIVITY 7-D 🍎

Using Strip Diagrams to Solve Proportion Problems

CCSS CCSS SMP2, SMP3, 6.RP.3

1. Spring Green paint is made by mixing blue paint with yellow paint in a ratio of 2 to 3. In parts (a), (b), and (c), explain how to reason about a strip diagram to make Spring Green paint.

Blue paint
Yellow paint

a. If you use 26 gallons of blue paint, how many gallons of yellow paint will you need?

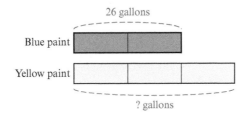

b. If you use 48 gallons of yellow paint, how many gallons of blue paint will you need? (Draw your own strip diagram!)

c. If you want to make 125 gallons of Spring Green paint, how many gallons of blue paint and how many gallons of yellow paint will you need?

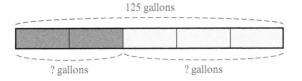

2. You are going to mix fruit juice and bubbly water in the ratio 5 to 3 to make your special punch. For each of the following, explain how to solve the problem by reasoning about a strip diagram.

   a. How much fruit juice and how much bubbly water will you need to make 32 cups of punch?

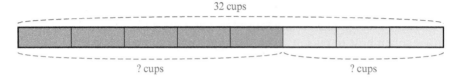

32 cups

? cups   ? cups

   b. How much fruit juice and how much bubbly water will you need to make 10 cups of punch? (Draw your own strip diagram!)

   c. How much fruit juice will you need if you will use 4 cups of bubbly water?

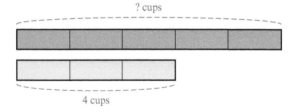

? cups

4 cups

   d. How much bubbly water will you need if you will use 4 cups of fruit juice? (Draw your own strip diagram!)

SECTION 7.2   CLASS ACTIVITY 7-E

Solving Proportion Problems by Reasoning about Multiplication and Division with Quantities

CCSS   CCSS SMP2, SMP3, 6.RP.3

Paint Problem: A paint company makes Peony Pink Paint by mixing red and white paint in the ratio 4 to 7. How many liters of white paint does the company need to mix with 35 liters of red paint to make Peony Pink Paint?

Solve the *Paint Problem* by reasoning about multiplication and division with quantities in as many ways as you can. In each case, describe the number of liters of white paint as a product $A \cdot B$, where A and B are suitable whole numbers, fractions, or mixed numbers that you derive from 4, 7, and 35. Attend carefully to our definition of multiplication. When you use division, explain what kind it is (how-many-groups or how-many-units-in-1-group). Use math drawings to support your explanations.

SECTION 7.2   CLASS ACTIVITY 7-F

Ratio Problem Solving with Strip Diagrams

CCSS   CCSS SMP1, SMP3, 7.RP.3

1. At lunch, there was a choice of pizza or a hot dog. Three times as many students chose pizza as chose hot dogs. All together, 160 students got lunch. How many students got pizza and how many got a hot dog? Draw a strip diagram to help you solve this problem. Explain your reasoning.

2. The ratio of Shauntay's cards to Jessica's cards is 5 to 3. After Shauntay gives Jessica 15 of her cards, both girls have the same number of cards. How many cards do Shauntay and Jessica each have now? Draw a strip diagram to help you solve this problem. Explain your reasoning.

3. The ratio of Shauntay's cards to Jessica's cards is 5 to 2. After Shauntay gives Jessica 12 of her cards, both girls have the same number of cards. How many cards do Shauntay and Jessica each have now? Draw a strip diagram to help you solve this problem. Explain your reasoning.

4. Make a new problem for your students by modifying part 2 or part 3. Change the ratio and change the number of cards that Shauntay gives to Jessica. When you make these changes, which ratios will make the problem easier, and which ratios will make it more difficult? Once you have chosen a ratio, can the number of cards that Shauntay gives to Jessica be any number, or do you need to take care in choosing this number? Explain.

SECTION 7.2   CLASS ACTIVITY 7-G

More Ratio Problem Solving

CCSS   CCSS SMP1, SMP2, 7.RP.3

1. Chandra made a milkshake by mixing $\frac{1}{2}$ cup of ice cream with $\frac{3}{4}$ cup of milk. Reason about quantities to determine how many cups of ice cream and milk Chandra should use if she wants to make the same milkshake (i.e., using the same ratio) with the following amounts:

   a. using 3 cups of ice cream

   b. to make 3 cups of milkshake

2. Russell was supposed to mix 3 tablespoons of weed killer concentrate with $1\frac{3}{4}$ cups of water to make a weed killer. By accident, Russell put in an extra tablespoon of weed killer concentrate, mixing 4 tablespoons of weed killer concentrate with $1\frac{3}{4}$ cups of water. How much water should Russell add to his mixture so that the ratio of weed killer concentrate to water will be correct? Reason about quantities to solve this problem.

SECTION 7.3 | CLASS ACTIVITY 7-H

How Can We Interpret and Use Values of a Ratio as Rates?

CCSS   CCSS SMP2, SMP8, 6.RP.2, 6.RP.3, 7.RP.1

1. A paint company makes a hue of orange paint by mixing red and yellow paint in the ratio 3 to 7. The company can make this hue of orange paint with different numbers of quarts of red and yellow paint, such as 10 quarts of red paint or 20 quarts of yellow paint.

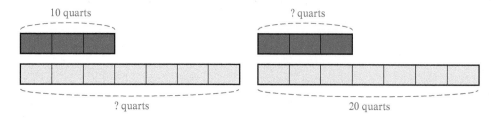

a. Explain how the values of the ratio $\frac{3}{7}$ and $\frac{7}{3}$ answer the following questions and that the answers don't depend on how many quarts of paint the company uses:

   • How many of the red strip does it take to make the yellow strip exactly?

   • How many of the yellow strip does it take to make the red strip exactly?

b. Use your answers to the questions in part (a) to help you find the unknown amounts that are indicated by question marks on the strip diagrams. Explain your reasoning.

c. In general, explain how to use values of the ratio to relate the number of quarts R of red paint and the number of quarts Y of yellow paint the company could use to make their orange paint.

   • Given a value for R, how could you find Y?

   • Given a value for Y, how could you find R?

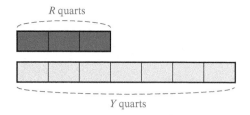

2. A race car is going 7 kilometers every 2 minutes and continues at that constant speed for a number of minutes.

 a. Put the values of the ratio $\frac{7}{2}$ and $\frac{2}{7}$ in the appropriate boxes on the double number lines and explain why they go there.

 Then use those values to help you find the unknown amounts that are indicated by question marks on the double number lines. Explain your reasoning.

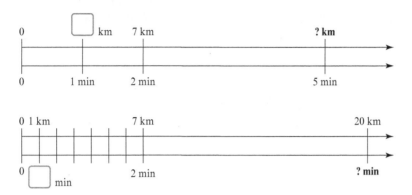

 b. In general, explain how to use values of the ratio to relate the number of minutes T and the number of kilometers D that the race car has driven.

 • Given a value for T, how could you find D?

 • Given a value for D, how could you find T?

SECTION 7.3 **CLASS ACTIVITY 7-I**

Solving Proportions by Cross-Multiplying Fractions

CCSS CCSS SMP3

Recipe Problem A recipe that serves 6 people calls for $2\frac{1}{2}$ cups of flour. How much flour will you need to serve 10 people, assuming that the ratio of people to cups of flour remains the same?

One familiar way to solve this problem is by letting x be the number of cups of flour we need to serve 10 people and setting two fractions equal to each other:

$$\frac{x}{10} = \frac{2\frac{1}{2}}{6}$$

Next, we cross-multiply to obtain the equation

$$6 \cdot x = 10 \cdot 2\frac{1}{2}$$

Finally, we solve for x by dividing both sides of the equation by 6. Therefore,

$$x = \frac{10 \cdot 2\frac{1}{2}}{6} = \frac{10 \cdot \frac{5}{2}}{6} = \frac{25}{6} = 4\frac{1}{6}$$

We need $4\frac{1}{6}$ cups of flour to serve 10 people.

Let's investigate the rationale for this method of solving proportions.

1. Interpret the meaning of the fractions

$$\frac{x}{10} \quad \text{and} \quad \frac{2\frac{1}{2}}{6}$$

   in terms of the recipe problem. (Remember that we can interpret fractions in terms of division.) Explain why these two fractions should be equal.

2. What is the rationale behind the procedure of cross-multiplying?

3. We could have set the equation up as

$$\frac{10}{6} = \frac{x}{2\frac{1}{2}}$$

   Interpret the fractions $\frac{10}{6}$ and $\frac{x}{2\frac{1}{2}}$ in terms of the recipe problem. Why should they be equal?

SECTION 7.4   CLASS ACTIVITY 7-J

Representing a Proportional Relationship with Equations

CCSS CCSS SMP2, 7.RP.2c

Suppose that Strip 1 and Strip 2 are stretchy, so they can get longer and shorter, but their lengths always remain in a 3 to 5 ratio. Strip 1 is X centimeters long, and Strip 2 is Y centimeters long, so X and Y vary together in a proportional relationship.

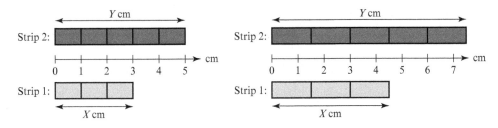

1. How many of Strip 1 does it take to make Strip 2? How many of Strip 2 does it take to make Strip 1? Do the answers depend on how many centimeters long the strips are? Explain!

2. If $X = 7$ cm, then what is Y? Explain. In general, given a value for X, how can you find the corresponding value of Y? Explain.

3. If $Y = 22$ cm, then what is X? Explain. In general, given a value for Y, how can you find the corresponding value of X? Explain.

4. Write and explain as many equations as you can that relate X and Y.

5. For each of the following, discuss why it is *not* a valid equation to relate X and Y.

   a. $Y = X + 2$

   b. $5Y = 3X$

   c. $5Y + 3X = $ combined length

6. If you did not do so already in part 4, use your answers to part 1 to derive and explain equations to relate X and Y.

SECTION 7.4 **CLASS ACTIVITY 7-K**

Relating Lengths and Heights of Ramps of the Same Steepness

CCSS CCSS SMP2, 7.RP.2

Imagine all the ramps that are X feet long and Y feet high where X and Y vary, but whose length and height are in the fixed 4-to-3 ratio. In the figures below, the 4 horizontal parts show the length of these ramps, and the 3 vertical parts show their height. Throughout this Class Activity, explain how to reason about these variable parts to solve problems about the ramps.

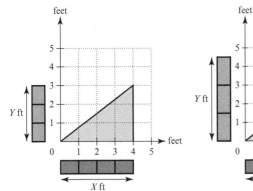 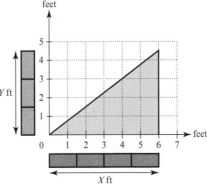 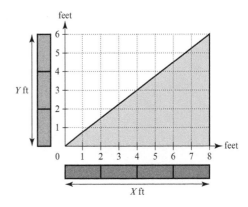

1. For these ramps, how many groups of their length amount (X feet) does it take to make their height amount (Y feet)? How many groups of their height amount does it take to make their length amount? Do the answers depend on the specific number of feet in their length and height? Explain!

2. For this kind of ramp, when $X = 10$ feet, what is Y?

3. For this kind of ramp, when $Y = 10$ feet, what is X?

4. By reasoning about the variable parts, develop and explain as many equations as you can that relate X and Y for these ramps.

5. If you did not do so already in part 4, use your answers to part 1 to derive and explain equations that relate X and Y.

6. Describe in as many ways as you can what is the same for all the ramps. Are there any equations or parts of equations that can help you describe or quantify this sameness?

SECTION 7.4 **CLASS ACTIVITY 7-L**

Where Do Equations of Lines Come From?

CCSS CCSS SMP7, 8.EE.5, 8.EE.6

Consider the line that goes through the origin, $(0, 0)$, and the point $(2, 3)$. For all the points (X, Y) on this line, X and Y are in the ratio 2 to 3 (in Chapter 14 we will use similar triangles to explain why). Do the following during this Class Activity:

- Pretend that you do not already know about equations of lines.
- Explain how to reason from a variable-parts perspective by viewing the X- and Y-coordinates of points on the line as consisting of 2 parts and 3 parts, respectively, where the parts stretch and shrink as the point (X, Y) moves up and down the line.

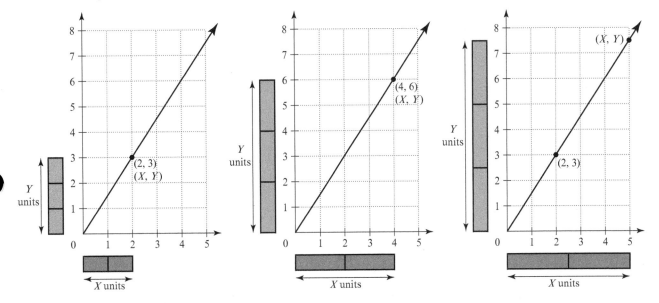

1. Explain how to reason about the variable parts to determine the Y-coordinate of the point on the line whose X-coordinate is 7.

2. Explain how to reason about the variable parts to determine the Y-coordinate of the point on the line whose X-coordinate is 5.

3. Explain how to reason about the variable parts to determine the X-coordinate of the point on the line whose Y-coordinate is 11.

4. By reasoning about the variable parts, develop and explain as many equations as you can to relate X and Y for points (X, Y) on the line.

5. Explain why for all points (X, Y) on the line, X and Y are related by the equations

$$Y = \frac{3}{2} \cdot X \quad \text{and} \quad X = \frac{2}{3} \cdot Y$$

In your explanation, take care to use our definition of multiplication and to interpret the meaning of $\frac{3}{2}$ and $\frac{2}{3}$ suitably. (Your explanation need only pertain to nonnegative values of X and Y.)

SECTION 7.4   CLASS ACTIVITY 7-M

Comparing Tables, Graphs, and Equations

CCSS   CCSS SMP2, 8.EE.5

The tables below show total distances and elapsed times as Devonte, Kellie, and Heather walked along a running track.

Devonte						
Meters	3	6	9	12	15	18
Seconds	2	4	6	8	10	12

Kellie					
Meters		4	8	12	16
Seconds		3	6	9	12

Heather						
Meters	1	2	4	7	11	16
Seconds	2	4	6	8	10	12

1. Sketch graphs based on the three tables. Compare and contrast the graphs and what they show about how the three students walked. Also, discuss why it makes sense to connect the points and to include the point $(0, 0)$.

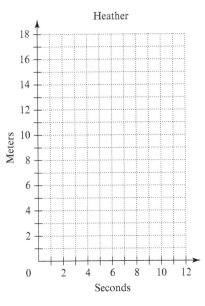

Heather

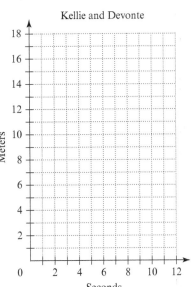

Kellie and Devonte

2. Which students are walking at a constant speed? How can you tell?

3. Who walked faster, Kellie or Devonte? Explain in several different ways how you can tell from their graphs.

4. A fourth student, Ashga, also walked along the running track with total distances and elapsed times shown below. Who walked faster, Ashga, Kellie, or Devonte? Explain how you can tell in several different ways.

Ashga			
Meters	11	22	33
Seconds	7	14	28

5. A fifth student, Brittany, walked along the track so that her total distance, d, in meters, and elapsed time, t, in seconds are given by the equation

$$d = \frac{9}{5} \cdot t$$

How does Brittany's walking compare to the others? Explain!

SECTION 7.5   CLASS ACTIVITY 7-N

How Are the Quantities Related?

CCSS   CCSS SMP3, 7.RP.2a

Use our definition of multiplication throughout this activity. When you use division, say what kind it is.

1. A painting company will send a crew of painters to paint a large wall. All painters work at the same steady pace. If a crew of 6 painters does the job, they will each paint an area of 300 square meters.

   a. If a crew of 18 painters does the job, how much area will each painter paint? Make a math drawing to help you explain.

   b. If a crew of 10 painters does the job, how much area will each painter paint?

   c. If each painter on the crew will paint 200 square meters, how many painters will be on the crew?

   d. Suppose X painters will paint Y square meters each to complete the job, where X and Y are unspecified numbers, which can vary. Find and explain an equation that relates 6, 300, X, and Y.

   e. Are X and Y in a proportional relationship? What are several different ways you can tell?

2. A restaurant will hire a crew of cooks to stuff dumplings for a large party. All the cooks stuff dumplings at the same steady pace, and they all work at the same time. If a crew of 8 cooks does the job, it will take 6 hours to stuff all the dumplings.

   a. If a crew of 2 cooks does the job, how long will it take? Make a math drawing to help you explain.

   b. If a crew of 3 cooks does the job, how long will it take?

   c. If the job needs to be done in 4 hours, how many cooks should be on the crew?

   d. Suppose X cooks will work for Y hours to do the job, where X and Y are unspecified numbers, which can vary. Find and explain an equation that relates 8, 6, X, and Y.

   e. Are X and Y in a proportional relationship? What are several different ways you can tell?

SECTION 7.5 CLASS ACTIVITY 7-O

Can You Use a Proportion or Not?

CCSS   CCSS SMP2, SMP4, 7.RP.2a

1. Ken used 3 loads of stone pavers to make a 10-foot-by-10-foot square patio. Ken wants to make another square patio, this one 20 feet by 20 feet, so he sets up the proportion

$$\frac{3 \text{ loads}}{10 \text{ feet}} = \frac{x \text{ loads}}{20 \text{ feet}}$$

Is this correct? If not, why not? Is there another way that Ken could solve the problem?

2. In a cookie factory, 4 assembly lines make enough boxes of cookies to fill a truck in 10 hours. How long will it take to fill the truck if 8 assembly lines are used? Is the proportion

$$\frac{10 \text{ hours}}{4 \text{ lines}} = \frac{x \text{ hours}}{8 \text{ lines}}$$

appropriate for this situation? Why or why not? If not, can you solve the problem another way? (Assume that all assembly lines work at the same steady rate.)

3. In the cookie factory of part 2, how long will it take to fill a truck if 6 assembly lines are used? (If you get stuck here, move on to the next problem and come back.)

4. Robyn used the following reasoning to solve the previous problem:

   "Four assembly lines fill a truck in 10 hours, so 8 assembly lines should fill a truck in half that time, so, in 5 hours. Since 6 assembly lines is halfway between 4 and 8, it ought to take halfway between 10 hours and 5 hours, or $7\frac{1}{2}$ hours, to fill a truck."

   Robyn's reasoning seems quite reasonable, but is it really correct? Let's look carefully. Fill in the following table by thinking logically about the assembly lines:

Number of hours			10			
Number of lines	1	2	4	8	16	32

   Now apply Robyn's reasoning again, but to 1 assembly line versus 32. Sixteen assembly lines is approximately halfway between 1 and 32. But is the corresponding number of hours also approximately halfway between?

   What can you conclude about Robyn's reasoning?

A Proportional Relationship versus an Inversely Proportional Relationship

CCSS CCSS SMP2, SMP4, 7.RP.2a, 8.F.3

1. At a bakery, 2 people can frost a total of 50 cupcakes in 12 minutes. Assume that all people work together at the same time and at the same steady rate.

   a. Make math drawings to help you explain how to fill in the blanks in part (a). Fill in the tables and answer the questions. Then compare and contrast the two relationships.

Relationship: Number of people ←→ Number of cupcakes when working for 12 minutes.	Relationship: Number of people ←→ Number of minutes when frosting 50 cupcakes.
(a) 2 times as many people frost _____ as many cupcakes. $\frac{1}{2}$ as many people frost _____ as many cupcakes. N times as many people frost _____ as many cupcakes.	(a) 2 times as many people take _____ as long. $\frac{1}{2}$ as many people take _____ as long. N times as many people take _____ as long.

(b) Cupcakes	50							
People	1	2	3	4	5	6	7	8

(b) Minutes	12							
People	1	2	3	4	5	6	7	8

(c) Find a • or ÷ relationship between number of people, number of cupcakes:	(c) Find a • or ÷ relationship between number of people, number of minutes:

(d) What type of relationship is it between number of people, number of cupcakes? How can you tell?	(d) What type of relationship is it between number of people, number of minutes? How can you tell?

   b. On separate paper, graph the points in the tables. Use the horizontal x-axis for the number of people. How are the shapes of the graphs different?

2. For each of the following, determine if the relationship between X and Y is proportional, inversely proportional, or neither. Explain!

   a. There is a large pile of sand that needs to be hauled away. X trucks will each take Y loads to haul away the sand.

   b. There are 20 trucks. Of those trucks, X trucks will haul sand, and the rest, Y trucks, will haul gravel.

   c. There are 20 trucks. Each truck will haul X loads of sand. All loads are the same size. All together, the trucks will haul Y tons of sand.

   d. There is a large sand pile. After X identical truck-loads of sand have been hauled away, there are Y tons of sand left in the pile.

How Should We Describe the Change?

CCSS   CCSS SMP2, 4.OA.2, 7.RP.3

A store raised some of its prices:

- A carton of milk went from $2 to $3.
- A box of laundry detergent went from $5 to $6.
- A small tube of makeup went from $10 to $15.
- A large tube of makeup went from $20 to $30.

The milk and the laundry detergent each went up by $1. But does that $1 increase seem equally significant in both cases?

The small tube of makeup went up by $5 and the large tube went up by $10. Does that mean that the price of the large tube of makeup went up more? Discuss!

SECTION 7.6   CLASS ACTIVITY 7-R

Calculating Percent Increase and Decrease

CCSS CCSS SMP3, 7.RP.3, 7.EE.2

1. Brand A cereal used to be sold in a 20-ounce box. Now Brand A cereal is sold in a 23-ounce box.

   a. Calculate the increase in the weight of cereal in a Brand A box as a percentage of the original weight.

   b. Now calculate the new weight of a Brand A box of cereal as a percentage of the original weight, and subtract 100%.

   c. Why does it make sense that the calculations in (a) and (b) produce the same results?

2. There were 20 gallons of gas in a tank. Now there are only 15 gallons left.

   a. Calculate the decrease in the amount of gas in the tank as a percentage of the original amount.

   b. Now calculate the new amount of gas in the tank as a percentage of the original amount, and subtract it from 100%.

   c. Why does it make sense that the calculations in (a) and (b) produce the same results?

SECTION 7.6   CLASS ACTIVITY 7-S 🏺

Calculating Amounts from a Percent Increase or Decrease

CCSS CCSS SMP3, 7.RP.3, 7.EE.2

1. The price of a Loungy Chair was $400. The price of this chair has just gone up by 20%.

   Complete the percent table (fill in steps as needed) and explain why the blank must be the new price of the Loungy Chair.

   $$100\% \rightarrow \$400$$

   $$120\% \rightarrow \underline{\hspace{2cm}}$$

2. A set of sheets was $60. The sheets are now on sale for 15% off.

   Complete the percent table (fill in steps as needed) and explain why the blank must be the new price of the sheets. Where does the 85% in the percent table come from?

   $$100\% \rightarrow \$60$$

   $$85\% \rightarrow \underline{\hspace{2cm}}$$

3. The price of a suit just went up by 20%. The new price, after the increase, is $180.
   a. Complete the percent table and explain why the blank will be the price of the suit before the increase.

   $$120\% \rightarrow \$180$$

   $$100\% \rightarrow \underline{\hspace{2cm}}$$

   Where does the 120% come from, and why does it correspond to $180?

   b. Explain why you *can't* calculate the price of the suit before the increase by decreasing $180 by 20%.

4. The price of a sofa went down by 20%. The new reduced price is $400.
   a. Complete the percent table and explain why the blank will be the price of the sofa before the reduction.

   $$80\% \rightarrow \$400$$

   $$100\% \rightarrow \underline{\hspace{2cm}}$$

   Where does the 80% come from, and why does it correspond to $400?

   b. Explain why you *can't* calculate the price of the sofa before the reduction by increasing $400 by 20%.

SECTION 7.6   CLASS ACTIVITY 7-T

Can We Solve It This Way?

CCSS   CCSS SMP3

1. The price of a cruise increased by 15%. The new price is $2300. What was the price before the increase?

   Here is how Matt solved the problem:

   First I found 10% and that's $230. Then 5% is half of that, so $115. So 15% is $345. So I took $345 away from $2300, which leaves $1955, and that's the answer.

   Discuss Matt's method. Is it correct or not? Show how to solve the problem in another way.

2. Whoopiedoo makeup used to be sold in 4-ounce tubes. Now it's sold in 5-ounce tubes for the same price. Ashlee says the label should read "25% more," whereas Carolyn thinks it should read "20% more." Who is right, who is wrong, and why?

3. The Film Club increased from 15 members to 45 members. Amy says that's a 300% increase. Kaia says it's a 200% increase. Who is right, who is wrong, and why?

SECTION 7.6   CLASS ACTIVITY 7-U

Percent Problem Solving

CCSS   CCSS SMP1

Strip diagrams may help you solve some of these problems.

1. At first, Prarie had 10% more than the cost of a computer game. After Prarie spent $7.50, they had 15% less than the cost of the computer game. How much did the computer game cost? How much money did Prarie have at first? Explain your reasoning.

2. One mouse weighs 20% more than another mouse. Together, the two mice weigh 66 grams. How much does each mouse weigh? Explain your reasoning.

3. There are two vats of orange juice. After 10% of the orange juice in the first vat is poured into the second vat, the first vat has 3 times as much orange juice as the second vat. By what percent did the amount of juice in the second vat increase when the juice from the first vat was poured into it? Explain your reasoning.

Percent Change and the Commutative Property of Multiplication

CCSS   CCSS SMP3, 7.EE.2

Which, if either, of the following two options will result in the lower price for a pair of pants?

- The price of the pants is marked up by 10% and then marked down by 20% from the increased price.
- The price of the pants is marked down by 20% and then marked up by 10% from the discounted price.

Both options involve marking up by 10% and marking down by 20%. The difference is the order in which the marking up and marking down occur.

1. Before you do any calculations, make a guess about which of the two options should result in a lower price.

2. Suppose the pants cost $50 to start with. Which of the two options will result in a lower price?

3. How are

$$0.80 \cdot 1.10 \cdot 50 \quad \text{and} \quad 1.10 \cdot 0.80 \cdot 50$$

relevant to the question about the pants? How is the commutative property of multiplication relevant?

SECTION 8.1   CLASS ACTIVITY 8-A

Factors and Rectangles

CCSS   CCSS SMP8, 4.OA.4

1. Elsie has 24 square tiles that she wants to arrange in the shape of a rectangle in such a way that the rectangle is completely filled with tiles. What are the different rectangles that Elsie can make and what do they tell you about the factors of 24?

2. If Elsie has more than 24 square tiles, will she necessarily be able to make more rectangles than she could in part 1? Try some experiments. What does this tell you about factors?

Finding All Factors

CCSS   CCSS SMP3, 4.OA.4

1. Tyrese is looking for all the factors of 156. So far, Tyrese has divided 156 by all the counting numbers from 1 to 13, listing those numbers that divide 156 and listing the corresponding quotients. Here is Tyrese's work so far:

$$
\begin{array}{ll}
1, 156 & 1 \times 156 = 156 \\
2, 78 & 2 \times 78 = 156 \\
3, 52 & 3 \times 52 = 156 \\
4, 39 & 4 \times 39 = 156 \\
6, 26 & 6 \times 26 = 156 \\
12, 13 & 12 \times 13 = 156 \\
13, 12 & 13 \times 12 = 156 \\
\end{array}
$$

Should Tyrese keep checking to see if numbers larger than 13 divide 156, or can Tyrese stop dividing at this point? If so, why? What are all the factors of 156?

2. Find all the factors of 198 in an efficient way.

SECTION 8.1 **CLASS ACTIVITY 8-C**

Do Factors Always Come in Pairs?

CCSS CCSS SMP1

Carmina noticed that factors always seem to come in pairs. For example,

$$48 = 1 \times 48, 1 \text{ and } 48 \text{ are a pair of factors of } 48.$$
$$48 = 2 \times 24, 2 \text{ and } 24 \text{ are a pair of factors of } 48.$$
$$48 = 3 \times 16, 3 \text{ and } 16 \text{ are a pair of factors of } 48.$$
$$48 = 4 \times 12, 4 \text{ and } 12 \text{ are a pair of factors of } 48.$$
$$48 = 6 \times 8, 6 \text{ and } 8 \text{ are a pair of factors of } 48.$$

The number 48 has 10 factors that come in 5 pairs. Carmina wants to know if every counting number always has an even number of factors. Investigate Carmina's question carefully. When does a counting number have an even number of factors, and when does it not?

Why Can We Check the Ones Digit to Determine Whether a Number Is Even or Odd?

CCSS CCSS SMP3

Remember that a counting number is called *even* if that number of objects can be divided into groups of 2 with none left over:

Why is it valid to determine whether a number of objects can be divided into groups of 2 with none left over by checking the ones digit of the number? We will investigate this question in several ways in this Class Activity.

1. What happens to the ones digits when we count by twos?

2. Recall that we can represent a whole number with base-ten bundles and think about putting a number of toothpicks into groups of 2 by working with such bundles. Why does it come down to the ones place to determine if a toothpick will be left over?

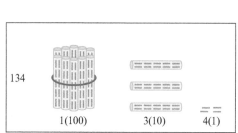
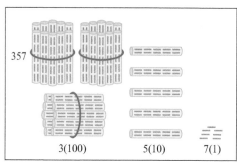

3. Working more generally, let *ABC* be a three-digit whole number with *A* hundreds, *B* tens, and *C* ones. Use the idea of representing *ABC* with base-ten bundles to help explain why *ABC* is divisible by 2 exactly when *C* is either 0, 2, 4, 6, or 8. What can we say about each of the *B* bundles of 10 toothpicks and each of the *A* bundles of 100 toothpicks when we divide the toothpicks into groups of 2?

SECTION 8.2 CLASS ACTIVITY 8-E

Questions about Even and Odd Numbers

CCSS CCSS SMP3, SMP8

1. If you add an odd number and an odd number, what kind of number do you get? Investigate this question by working out examples. Then explain why your answer is always correct. Try to find several different explanations by working with the various equivalent ways of saying that a number is even or odd.

2. If you multiply an even number and an odd number, what kind of number do you get? Investigate this question by working out examples. Then explain why your answer is always correct. Try to find several different explanations by working with the various equivalent ways of saying that a number is even or odd.

Extending the Definitions of Even and Odd

CCSS CCSS SMP3, SMP6

We have defined even and odd only for counting numbers. What if we wanted to extend the definition of even and odd to other numbers?

1. If we extend the definitions of even and odd to all the integers, what should 0 be, even or odd? What should −5 be, even or odd? Explain.

2. Give definitions of even and odd that apply to all integers, not just to the counting numbers.

3. Would it make sense to extend the definitions of even and odd to fractions? Why or why not?

SECTION 8.3   **CLASS ACTIVITY 8-G** 🏆

The Divisibility Test for 3

CCSS CCSS SMP3, SMP8

1. Is it possible to tell if a counting number is divisible by 3 just by checking its last digit?

   Investigate this question by considering a number of examples. State your conclusion.

2. The divisibility test for 3 is this: Given a counting number, add its digits. If the sum is divisible by 3, then the original number is, too; if the sum is not divisible by 3, then the original number is not either.

   For each of the numbers listed, check that the divisibility test for 3 accurately predicts which numbers are divisible by 3.

   2570   14,928   11,111

3. Explain why the divisibility test for 3 is valid for three-digit counting numbers. In other words, explain why you can determine whether a three-digit counting number, *ABC*, is divisible by 3 by adding its digits, $A + B + C$, and determining if this sum is divisible by 3.

   To develop your explanation, consider the following:

   a. A counting number is divisible by 3 exactly when that many objects can be divided into groups of 3 with none left over.

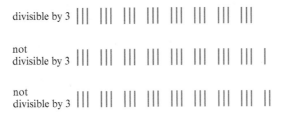

   b. Think about representing three-digit numbers with base-ten bundles.

   c. Think about dividing bundled toothpicks into groups of 3 by dividing *each individual bundle* of 10 and *each individual bundle* of 100 into groups of 3. How many toothpicks are left over from each bundle?

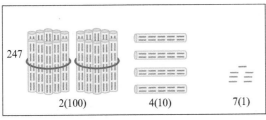

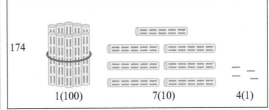

SECTION 8.4 | **CLASS ACTIVITY 8-H** 🏺

The Sieve of Eratosthenes

CCSS CCSS SMP3, 4.OA.4

1. Use the Sieve of Eratosthenes to find all the prime numbers up to 120. Start by circling 2 and crossing off every 2nd number after 2. Then circle 3 and cross off every 3rd number. (Cross off a number even if it already has been crossed off.) Continue in this manner, going back to the beginning of the list, circling the next number N that hasn't been crossed off, and then crossing off every Nth number until every number in the list is either circled or crossed off. The numbers that are circled at the end are the prime numbers from 2 to 120.

	2	3	4	5	6	7	8	9	10
11	12	13	14	15	16	17	18	19	20
21	22	23	24	25	26	27	28	29	30
31	32	33	34	35	36	37	38	39	40
41	42	43	44	45	46	47	48	49	50
51	52	53	54	55	56	57	58	59	60
61	62	63	64	65	66	67	68	69	70
71	72	73	74	75	76	77	78	79	80
81	82	83	84	85	86	87	88	89	90
91	92	93	94	95	96	97	98	99	100
101	102	103	104	105	106	107	108	109	110
111	112	113	114	115	116	117	118	119	120

2. Explain why the circled numbers must be prime numbers and why the numbers that are crossed off are not prime numbers.

SECTION 8.4 CLASS ACTIVITY 8-I

The Trial Division Method for Determining Whether a Number Is Prime

CCSS CCSS SMP3, 4.OA.4

Materials You will need the list of prime numbers you found in the previous Class Activity.

1. Using the trial division method and your list of primes, determine whether the three numbers listed are prime numbers. Record the results of your trial divisions below the number. (The first few are done for you.) You will need these results for part 3.

   239
   $239 \div 2 = 119.5$
   $239 \div 3 = 79.67 \ldots$

   323
   $323 \div 2 = 161.5$
   $323 \div 3 = 107.67 \ldots$

   4001
   $4001 \div 2 = 2000.5$
   $4001 \div 3 = 1333.67 \ldots$

2. How do you know when to stop with the trial division method? To help you answer, look at the list of divisions you did in part 1. As you go down each list, what happens to the divisor and the quotient? If your number *was* divisible by some whole number (other than 1), at what point would that whole number be known?

3. In the trial division method, you determine only whether your number is divisible by *prime* numbers. Why is this legitimate? Why don't you also have to find out if your number is divisible by other numbers such as 4, 6, 8, 9, 10, and so on?

SECTION 8.4   CLASS ACTIVITY 8-J

Do Different Factor Trees Produce Different Results?

CCSS   CCSS SMP3

1. Make a factor tree for 240 and use it to write 240 as a product of prime numbers.

2. Compare your factor tree for 240 to a classmate's. Are the factor trees identical in all respects? If so, make a different factor tree. Do the different factor trees produce the same end result?

3. Make a factor tree for 180 and use it to write 180 as a product of prime numbers.

4. Compare your factor tree for 180 to a classmate's. If they are identical, make a different factor tree. Do the different factor trees produce the same end result?

5. Will it be obvious to students that the end result of factoring a number into a product of prime numbers must be the same, no matter how they make their factor tree? Discuss!

SECTION 8.4 CLASS ACTIVITY 8-K

Problem Solving with Products of Prime Numbers

1. Lindsay factored 637 as $637 = 7 \times 7 \times 13$. When Lindsay was then asked to factor 637^2 as a product of prime numbers, she first multiplied $637 \times 637 = 405{,}769$; then she divided 405,769 by 2, then by 3, then by 5, and so on, in order to factor 405,769. Is there an easier way for Lindsay to factor 637^2 into a product of prime numbers? Explain.

2. Given that $527 = 17 \times 31$, is there a way to check if 77 is a factor of 527 without actually dividing 527 by 77? Explain.

3. Given that $625 = 5^4$, find all the factors of 625. Explain how you know you have found them all.

   If you are considering using the general method for finding factors from Section 8.1, see if you can find a more efficient method that relies on the factorization you are given!

4. Given that $2431 = 11 \times 13 \times 17$, find all the factors of 2431. Explain how you know you have found them all.

SECTION 8.5 | CLASS ACTIVITY 8-L

How Do Multiples Align?

Materials Optionally, use Download 8-1 at bit.ly/2SWWFUX, markers, and scissors. If possible, copy the Download onto transparencies, and if so, use markers in light colors.

If you are using the Download: Cut the Download into 8 separate numbered strips (each participant will need only 4 of them). Then in parts 1 – 3 below, line up or overlay the relevant strips instead of coloring the strips below.

- Color the multiples of 4 on one strip.
- Color the multiples of 6 on a second strip.
- Color the multiples of 9 on a third strip.
- Color the multiples of 10 on a fourth strip.

1. Color the multiples of 4 on the first strip and the multiples of 6 on the second strip. What do you notice? When are the same numbers colored on both strips?

| 1 | 2 | 3 | 4 | 5 | 6 | 7 | 8 | 9 | 10 | 11 | 12 | 13 | 14 | 15 | 16 | 17 | 18 | 19 | 20 | 21 | 22 | 23 | 24 | 25 | 26 | 27 | 28 | 29 | 30 | 31 | 32 | 33 | 34 | 35 | 36 | 37 | 38 | 39 | 40 |

2. Color the multiples of 6 on the first strip and the multiples of 9 on the second strip. What do you notice? When are the same numbers colored on both strips?

| 1 | 2 | 3 | 4 | 5 | 6 | 7 | 8 | 9 | 10 | 11 | 12 | 13 | 14 | 15 | 16 | 17 | 18 | 19 | 20 | 21 | 22 | 23 | 24 | 25 | 26 | 27 | 28 | 29 | 30 | 31 | 32 | 33 | 34 | 35 | 36 | 37 | 38 | 39 | 40 |

3. Color the multiples of 4 on the first strip and the multiples of 10 on the second strip. What do you notice? When are the same numbers colored on both strips?

| 1 | 2 | 3 | 4 | 5 | 6 | 7 | 8 | 9 | 10 | 11 | 12 | 13 | 14 | 15 | 16 | 17 | 18 | 19 | 20 | 21 | 22 | 23 | 24 | 25 | 26 | 27 | 28 | 29 | 30 | 31 | 32 | 33 | 34 | 35 | 36 | 37 | 38 | 39 | 40 |

SECTION 8.5 CLASS ACTIVITY 8-M

The Slide Method

CCSS   CCSS SMP8

1. Examine the initial and final steps of a "slide" that was used to find the GCF and LCM of 900 and 360. Try to determine how it was made. Then make another slide to find the GCF and LCM of 900 and 360.

A Slide

initially: | 900 | 360 |

final:

10	900	360
2	90	36
3	45	18
3	15	6
	5	2

$GCF = 10 \cdot 2 \cdot 3 \cdot 3 = 180$

$LCM = 10 \cdot 2 \cdot 3 \cdot 3 \cdot 5 \cdot 2 = 1800$

2. Use the slide method to find the GCF and LCM of 1080 and 1200 and to find the GCF and LCM of 675 and 1125.

3. Why does the slide method work?

SECTION 8.5   **CLASS ACTIVITY 8-N**

Construct Arguments and Critique Reasoning about GCFs and LCMs

CCSS   CCSS SMP3

1. Find the GCF and LCM of $2^6 \cdot 3^4 \cdot 7^3$ and $2^5 \cdot 3^{10} \cdot 7^5$ and explain your reasoning.

2. Some students are finding GCFs and LCMs of numbers written as products of powers of primes.

   • Knowshon says he has a way to find the GCF and LCM: For the GCF, take the smallest power of each prime; for the LCM take the largest power of each prime.

   • Abbey is wondering if Knowshon's method is backward. She asks: Why don't you take the greatest power for the greatest common factor?

   • Matt says that Knowshon's method won't work for numbers like $2^7 \cdot 3^8 \cdot 5^3$ and $2^9 \cdot 3^4 \cdot 7$, because they don't both have 5s and 7s and also the 7 doesn't have an exponent.

   Discuss the students' ideas and objections. (Recall that any nonzero number raised to the 0 power is 1.)

SECTION 8.5 CLASS ACTIVITY 8-O

Model with GCFs and LCMs

CCSS CCSS SMP4

Solve each problem and explain your solution. Say whether the problem involves the GCF or the LCM.

1. Pencils come in packages of 18; erasers that fit on top of these pencils come in packages of 24. What is the smallest number of pencils and erasers that you can buy so that all the pencils and erasers can be paired? (Assume that you can't buy partial packages.)

2. A class is clapping and snapping to a steady beat. Half of the class uses the pattern

   snap, snap, clap, snap, snap, clap, . . .

   The other half of the class uses the pattern

   snap, clap, snap, clap, snap, clap, . . .

   When will the whole class be clapping together?

3. Mary will make a small 8-inch-by-12-inch rectangular quilt for a doll house out of identical square patches. Each square patch must have side lengths that are a whole number of inches, and no partial squares are allowed in the quilt. Other than using 1-inch-by-1-inch squares, what size squares can Mary use to make her quilt? Show Mary's other options below. What are the largest squares that Mary can use?

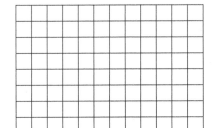 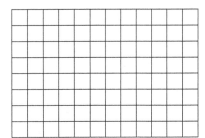

4. Two gears are meshed, as shown in the figure below, with the stars on each gear aligned. The large gear has 36 teeth, and the small gear has 15 teeth. Each gear rotates around a pin through its center. How many revolutions will the large gear have to make and how many revolutions will the small gear make in order for the stars to be aligned again?

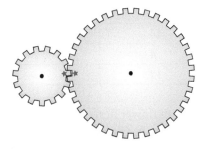

Spirograph Flower Designs

CCSS   CCSS SMP1

Each flower design below is created by starting at a dot, and connecting each subsequent *N*th dot until returning to the starting dot.

Design 1:
36 dots; a petal connects
every 8th dot, 9 petals

Design 2:
36 dots; a petal connects
every 15th dot, 12 petals

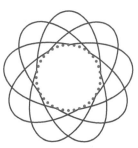

Design 3:
36 dots; a petal connects
every 16th dot, 9 petals

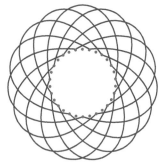

Design 4:
30 dots; a petal connects
every 14th dot, 15 petals

Design 5:
30 dots; a petal connects
every 4th dot, ____ petals

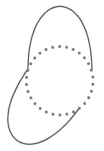

Design 6:
30 dots; a petal connects
every 12th dot, ____ petals

1. For the first four flower designs, find a relationship between the number of dots at the center of the flower, the way the petals were made, and the number of petals in the flower design. These numbers are listed in the following table:

Design	Number of Dots	A Petal Connects Every ____ th Dot	Number of Petals
1	36	8	9
2	36	15	12
3	36	16	9
4	30	14	15
5	30	4	
6	30	12	

2. Predict the number of petals that the 5th and 6th flower designs will have. Then complete the designs to see if your prediction was correct. (To complete the designs, you might find it easiest to count dots forward and then draw the petal backward.) See if you can explain why the numbers are related the way they are!

SECTION 8.6 CLASS ACTIVITY 8-Q 🝊

Decimal Representations of Fractions

CCSS CCSS SMP8, 7.NS.2d, 8.NS.1

1. The decimal representations of the following fractions are shown to 16 decimal places, with no rounding:

$$\frac{1}{12} = 0.0833333333333333\ldots \qquad\qquad \frac{1}{4} = 0.25$$

$$\frac{1}{11} = 0.9090909090909090\ldots \qquad\qquad \frac{2}{5} = 0.4$$

$$\frac{113}{33} = 3.4242424242424242\ldots \qquad\qquad \frac{37}{8} = 4.625$$

$$\frac{491}{550} = 0.8927272727272727\ldots \qquad\qquad \frac{17}{50} = 0.34$$

$$\frac{14}{37} = 0.3783783783783783\ldots \qquad\qquad \frac{1}{125} = 0.008$$

$$\frac{35}{101} = 0.3465346534653465\ldots \qquad\qquad \frac{9}{20} = 0.45$$

In what way are the decimal representations in the first column similar? In what way are the decimal representations of the fractions in the second column similar?

2. Complete the next set of calculations, using the standard division algorithm (not a calculator!) to find the decimal representations of $\frac{4}{7}$ and $\frac{3}{8}$. At each step in the long-division process, write down the remainder you obtain.

```
     0.5                                          0.
  7)4.0000000    remainder 4                    8)3.0000000    remainder 3
   -35                                            ___
     5            remainder 5                                  remainder ___

   ___            remainder___                    ___          remainder___
   ___            remainder___                    ___          remainder___
   ___            remainder___                    ___          remainder___
   ___            remainder___                    ___          remainder___
   ___            remainder___                    ___          remainder___
   ___            remainder___                    ___          remainder___
```

- What happened to the decimal representation of the fraction when you got a remainder that you had before?

- What happened to the decimal representation of the fraction when you got a remainder of 0?

3. Without actually carrying it out, imagine doing division to find the decimal representation of $\frac{7}{31}$.

   a. What remainders could you possibly get in the division process when finding $7 \div 31$? For example, could you possibly get a remainder of 45 or 73 or 32? How many different remainders are theoretically possible?

   b. If you were doing division to find $7 \div 31$ and you got a remainder of 0 somewhere along the way, what would that tell you about the decimal representation of $\frac{7}{31}$?

   c. If you were doing division to find $7 \div 31$ and you got a remainder you had gotten before, what would then happen in the decimal representation of $\frac{7}{31}$?

   d. Now use your answer to parts (a), (b), and (c) to explain why the decimal representation of $\frac{7}{31}$ must either terminate or eventually repeat after at most 30 decimal places.

4. In general, suppose that $\frac{A}{B}$ is a fraction, where A and B are whole numbers and A is less than B. Explain why the decimal representation of $\frac{A}{B}$ must either terminate or begin to repeat after at most $B - 1$ decimal places.

5. Could the following number be the decimal representation of a fraction?

   $$0.101001000100001000001000000 1 \ldots$$

   The decimal representation continues forever with the pattern of more and more 0s in between 1s. Explain your answer.

SECTION 8.6 CLASS ACTIVITY 8-R

Writing Terminating and Repeating Decimals as Fractions

CCSS CCSS SMP8, 8.NS.1

1. By using denominators that are suitable powers of 10, show how to write the following terminating decimals as fractions:

$0.137 =$ $0.25567 =$

$13.89 =$ $329.2 =$

2. Write the following fractions as decimals, and observe the pattern:

$$\frac{1}{9} =$$

$$\frac{1}{99} =$$

$$\frac{1}{999} =$$

$$\frac{1}{9999} =$$

$$\frac{1}{99,999} =$$

3. Using the decimal representations of $\frac{1}{9}, \frac{1}{99}, \ldots$, that you found in part 2, show how to write the following decimals as fractions:

$0.\overline{2} = 0.222222\ldots =$ $00.\overline{08} = 0.080808\ldots =$

$0.\overline{003} = 0.003003\ldots =$ $0.\overline{52} = 0.525252\ldots =$

$0.\overline{1234} =$ $0.\overline{123456} =$

4. Use the fact that $0.\overline{49} = \frac{49}{99}$ to write the next four repeating decimals as fractions.
Hint: Shift the decimal point by dividing by suitable powers of 10.

$0.0\overline{49} =$ $0.00\overline{49} =$

$0.000\overline{49} =$ $0.0000\overline{49} =$

5. Use the results of part 4, together with facts such as

$$0.3\overline{49} = 0.3 + 0.0\overline{49}$$

to write the following repeating decimals as fractions:

$7.3\overline{49} =$ $0.12\overline{49} =$

$1.2\overline{49} =$ $0.111\overline{49} =$

What Is 0.9999 . . . ?

CCSS CCSS SMP3, 8.NS.1

1. Use the fact that $\frac{1}{9} = 0.\overline{1} = 0.111111111\ldots$ to determine the decimal representations of the following fractions:

$$\frac{2}{9} = \qquad\qquad \frac{3}{9} = \qquad\qquad \frac{4}{9} = \qquad\qquad \frac{5}{9} =$$

$$\frac{6}{9} = \qquad\qquad \frac{7}{9} = \qquad\qquad \frac{8}{9} = \qquad\qquad \frac{9}{9} =$$

What can you conclude about $0.\overline{9}$?

2. Add longhand:

$$0.9999999999\ldots$$
$$+\ 0.1111111111\ldots$$

Note that the nines and ones
repeat forever.

Now subtract longhand: $\qquad -0.1111111111\ldots$

Look back at what you just did: Starting with $0.\overline{9}$, you added and then subtracted $0.\overline{1}$. What does this tell you about $0.\overline{9}$?

3. Let N stand for the number $0.\overline{9} = 0.9999999999\ldots$, so $N = 0.9999999999\ldots$

Below, write the decimal representation of $10N$ and then subtract N from $10N$ in two ways—in terms of N and as decimals.

$$\begin{array}{cc} \text{In terms of } N: & \text{As decimals:} \\ 10N & \\ \underline{-N} & \underline{-0.9999999999\ldots} \end{array}$$

What can you conclude about $0.\overline{9}$?

4. Given that the number 1 has two different decimal representations, namely, 1 and $0.\overline{9}$, find different decimal representations of the following numbers:

$$17 = \qquad\qquad 23.42 = \qquad\qquad 139.8 =$$

SECTION 8.6 **CLASS ACTIVITY 8-T**

The Square Root of 2

CCSS CCSS SMP3

1. If the sides of a square are 1 unit long, then how long is the diagonal of the square?

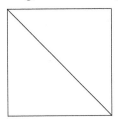

2. Use a calculator to find the decimal representation of $\sqrt{2}$. Based on your calculator's display, does it look like $\sqrt{2}$ is rational or irrational? Why? Can you tell *for sure* just by looking at your calculator's display?

3. Find the decimal representation of

$$\frac{1,414,213,562}{999,999,999}$$

Is this number rational or irrational? Compare with part 2.

4. Suppose that it were somehow possible to write the square root of 2 as a fraction $\frac{A}{B}$, where A and B are counting numbers:

$$\sqrt{2} = \frac{A}{B}$$

Show that, in this case, we would get the equation

$$A^2 = 2 \cdot B^2$$

5. Suppose A is a counting number, and imagine factoring it into a product of prime numbers. For example, if A is 30, then you factor it as

$$A = 2 \cdot 3 \cdot 5$$

Now think about factoring A^2 as a product of prime numbers. For example, if $A = 30$, then

$$A^2 = 2 \cdot 3 \cdot 5 \cdot 2 \cdot 3 \cdot 5$$

Could A^2 have an odd number of prime factors? Make a general qualitative statement about the number of prime factors that A^2 has.

6. Now suppose that B is a counting number, and imagine factoring the number $2 \cdot B^2$ into a product of prime numbers. For example, if $B = 15$, then

$$2 \cdot B^2 = 2 \cdot 3 \cdot 5 \cdot 3 \cdot 5$$

Could $2B^2$ have an even number of prime factors? Make a general qualitative statement about the number of prime factors that $2 \cdot B^2$ has.

7. Now use your answers in parts 5 and 6 to explain why a number in the form A^2 can never be equal to a number in the form $2 \cdot B^2$, when A and B are counting numbers.

8. What does part 7 lead you to conclude about the assumption in part 4 that it is somehow possible to write the square root of 2 as a fraction, where the numerator and denominator are counting numbers? Now what can you conclude about whether $\sqrt{2}$ is rational or irrational?

SECTION 9.1 **CLASS ACTIVITY 9-A** 🍎

Writing Expressions for Dot, Star, and Stick Designs

CCSS CCSS SMP7, 5.OA.1, 5.OA.2, 6.EE.1

Materials Optionally, use Downloads 9-1 and 9-2 at bit.ly/2SWWFUX to experiment and record your work.

1. For each flower design, write an expression for the total number of dots in the design. Each expression should involve both multiplication and addition. What do all the expressions have in common?

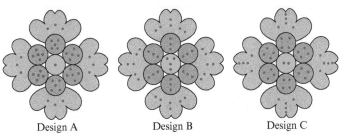

Design A   Design B   Design C

2. Write two different expressions for the total number of dots in the dot design below. Show or explain why your expressions work without evaluating them.

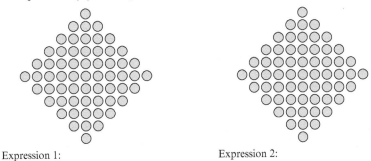

Expression 1:

_____

Expression 2:

_____

3. Write an expression for the total number of stars in the star design below. Show or explain why the other two expressions give the total number of stars without evaluating the expressions.

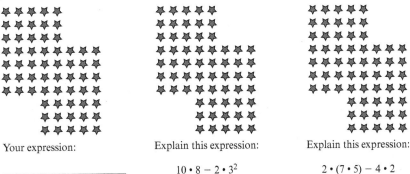

Your expression:

_____

Explain this expression:

$10 \cdot 8 - 2 \cdot 3^2$

Explain this expression:

$2 \cdot (7 \cdot 5) - 4 \cdot 2$

4. Write two different expressions for the total number of dots in the dot design below. Try to find an expression that involves subtraction and a power.

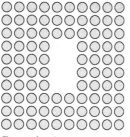 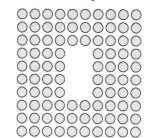

Expression 1:   Expression 2:

_____   _____

5. Write an expression for the total number of sticks in the stick design below.

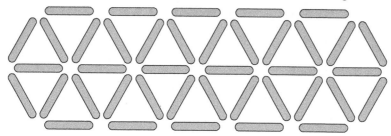

6. The stick design above is 6 sticks wide. How would you modify your expression in part 5 if the design were 8 sticks wide?

7. Make a dot or stick drawing for the expression

$$4^2 + 4^3$$

8. Make your own dot or stick drawing. Ask a partner to write an expression for the number of dots or sticks in your drawing.

SECTION 9.1 CLASS ACTIVITY 9-B

How Many High-Fives?

CCSS CCSS SMP1, SMP4

The *high-five problem* (traditionally known as the *handshake problem*): Twenty students are in a class. If every student high-fives with every other student, how many high-fives will there be?

1. Try to find two different expressions for the total number of high-fives among 20 students. One expression should involve addition, and the other expression should involve multiplication. Explain why each expression stands for the total number of high-fives.

2. Solve the high-five problem in part 1 by evaluating one of the expressions.

   Which expression is easiest to evaluate?

3. What if there were 50 students in the class? Write two expressions for the number of high-fives in this case; feel free to use an ellipsis (. . .) in one of your expressions! Explain why the expressions stand for the number of high-fives. Then solve the high-five problem in this case.

   Which expression is easiest to use to solve the high-five problem?

SECTION 9.1 **CLASS ACTIVITY 9-C**

Sums of Odd Numbers

CCSS CCSS SMP7, SMP8

Is there a quick way to add a bunch of consecutive odd numbers? This activity will help you find and explain another way to express a sum of odd numbers.

1. Calculate each of the next sums.

$$1 + 3 = \underline{\hspace{1cm}}$$
$$1 + 3 + 5 = \underline{\hspace{1cm}}$$
$$1 + 3 + 5 + 7 = \underline{\hspace{1cm}}$$
$$1 + 3 + 5 + 7 + 9 = \underline{\hspace{1cm}}$$
$$1 + 3 + 5 + 7 + 9 + 11 = \underline{\hspace{1cm}}$$
$$1 + 3 + 5 + 7 + 9 + 11 + 13 = \underline{\hspace{1cm}}$$

2. What is special about the solutions to the sums in part 1?

3. Based on your answer in part 2, predict the sum of the first 100 odd numbers.

4. Based on your answer in part 2, predict the next sum:
$$1 + 3 + 5 + 7 + 9 + \cdots + 91 + 93 + 95 + 97 + 99 = \underline{\hspace{1cm}}$$

5. Use the square designs below to explain why there are two ways to express sums of consecutive odd numbers.

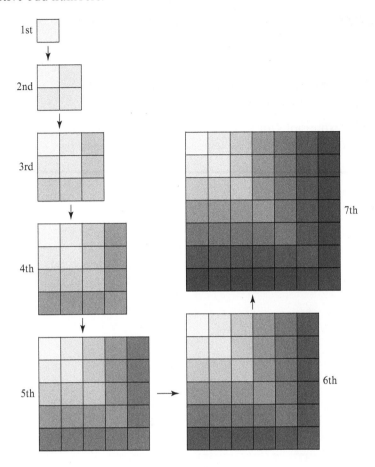

SECTION 9.1 **CLASS ACTIVITY 9-D**

Expressions with Fractions and Percent

CCSS CCSS SMP7

1. At a store, the price of an item is $300. After a month, the price is raised by 20%. After another month, the new price is raised by 25%.

   a. Write and explain two different expressions for the price of the item after the first month. Your expressions should involve 300 and 20. Include a math drawing as part of your explanation.

   b. Write and explain two different expressions for the price of the item after the second month. Your expressions should involve 300, 20, and 25. Include a math drawing as part of your explanation.

2. For each of the two rectangles below, write an expression using multiplication and addition (or subtraction) for the fraction of the area of the rectangle that is shaded. You may assume that parts that appear to be the same size really are the same size.

Rectangle 1

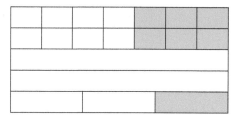

Rectangle 2

3. Shade

$$\frac{2}{3} \cdot \frac{4}{5} + \frac{1}{4} \cdot \frac{1}{10}$$

of a rectangle in such a way that you can tell the correct amount is shaded without evaluating the expression.

SECTION 9.1 **CLASS ACTIVITY 9-E**

Explain and Critique Evaluating Expressions with Fractions

CCSS CCSS SMP3

1. In order to evaluate

$$\frac{8}{35} \cdot \frac{35}{61}$$

we can cancel thus:

$$\frac{8}{35} \cdot \frac{35}{61} = \frac{8}{61}$$

Discuss the following equations and explain why they demonstrate that the canceling shown previously is legitimate:

$$\frac{8}{35} \cdot \frac{35}{61} = \frac{8 \cdot 35}{35 \cdot 61} = \frac{8 \cdot 35}{61 \cdot 35} = \frac{8}{61} \cdot \frac{35}{35} = \frac{8}{61}$$

2. Write equations to demonstrate that the canceling shown in the following equations is legitimate:

$$\frac{\overset{2}{\cancel{18}}}{5} \cdot \frac{7}{\underset{11}{\cancel{99}}} = \frac{2}{5} \cdot \frac{7}{11} = \frac{14}{55}$$

3. Which of the cancellations in parts (a) through (d) are correct, and which are incorrect? Explain your answers.

a. $\dfrac{\overset{6}{\cancel{36}} \cdot \overset{16}{\cancel{96}}}{\underset{1}{\cancel{6}}} = \dfrac{6 \cdot 16}{1} = 96$

b. $\dfrac{\overset{6}{\cancel{36}} \cdot 96}{\underset{1}{\cancel{6}}} = \dfrac{6 \cdot 96}{1} = 576$

c. $\dfrac{\overset{6}{\cancel{36}} + \overset{16}{\cancel{96}}}{\underset{1}{\cancel{6}}} = \dfrac{6 + 16}{1} = 22$

d. $\dfrac{\overset{6}{\cancel{36}} + 96}{\underset{1}{\cancel{6}}} = \dfrac{6 + 96}{1} = 102$

CLASS ACTIVITY 9-F

Equivalent Expressions

CCSS CCSS SMP7, 6.EE.3, 6.EE.4, 7.EE.1

1. There are J liters of juice in a vat. Someone pours $\frac{1}{3}$ of the juice out of the vat. Which of the next expressions give the number of liters of juice that remain in the vat? Explain.

   a. $J - \frac{1}{3}$ **b.** $(J - \frac{1}{3}) \cdot J$ **c.** $J - \frac{1}{3} \cdot J$ **d.** $J - \frac{1}{3}J$ **e.** $\frac{2}{3}J$

2. For each pair, determine if the expressions are equivalent or not. If they are equivalent, use properties of arithmetic to show why. If they are not, explain why not.

   a. $P - \frac{2}{5}P$ and $\frac{3}{5}P$

   b. $P - \frac{2}{5}$ and $\frac{3}{5}P$

   c. $7x - x$ and 7

   d. $7x - 7$ and x

   e. $3 \cdot (x \cdot y)$ and $(3 \cdot x) \cdot (3 \cdot y)$

   f. $3 \cdot (x + 2)$ and $3x + 6$

   g. $(x + y)^2$ and $x^2 + y^2$

   h. $x^4 y^4$ and $(xy)^8$

3. For each expression below, see if you can write an equivalent expression that is a product.

   a. $(x^2 + y^2) + (x^2 + y^2)z^2$

   b. $(x^2 + y^2) + (x^5 + y^5)$

   c. $x^7 y^3 z^4 + x^4 y^5 z^7$

SECTION 9.2 **CLASS ACTIVITY 9-G** 🍎

Expressions for Quantities

CCSS CCSS SMP2, SMP4, 6.EE.6, 7.EE.4

1. a. There are x tons of sand in a pile initially. Then $\frac{1}{4}$ of the sand in the pile is removed from the pile and, after that, another $\frac{2}{3}$ of a ton of sand is dumped onto the pile. Write two equivalent expressions in terms of x for the number of tons of sand that are in the pile now.

b. There are x tons of sand in a pile initially. Then $\frac{2}{3}$ of a ton of sand is dumped onto the pile and, after that, $\frac{1}{4}$ of the sand in the new, larger pile is removed. Write two equivalent expressions in terms of x for the number of tons of sand that are in the pile now.

c. Are your expressions in parts (a) and (b) equivalent or not? How can you tell?

2. For each of the following expressions, describe a corresponding situation. Be sure to say what x means in each situation.

a. $x - \frac{1}{4}x + 30$

b. $x - \frac{1}{4} + 30$

c. $(x + 30) - \frac{1}{4}(x + 30)$

d. $\frac{2}{3}(x - 60) + 20$

3. A T-shirt company has found that if it sells T-shirts for $x each, then it will sell $280 - 10x$ T-shirts per day. The company has daily fixed operating costs of $690. Each T-shirt costs the company $2 to make.

   a. Interpret the expression $280 - 10x$ in terms of the situation. What does it tell you?

   b. Write and explain an expression for the company's daily profit (income minus expenses) in terms of x.

4. At a store, the price of an item is $P. Consider three different scenarios:

   First scenario: Starting at the price $P, the price of the item was lowered by $A\%$. Then the new price was lowered by $B\%$.

   Second scenario: Starting at the price $P, the price of the item was lowered by $B\%$. Then the new price was lowered by $A\%$.

   Third scenario: Starting at the price $P, the price of the item was lowered by $(A + B)\%$.

   For each scenario, write an expression for the final price of the item. Are any of the expressions for the final prices equivalent?

Solving Equations by Reasoning about Expressions

CCSS   CCSS SMP7, 6.EE.5

Solve each of the equations by thinking about the expressions on both sides of the equal sign and reasoning about which value of the variable will make them equal. Do not use any standard algebraic techniques for solving equations that you may know. Explain your reasoning in each case.

1. $382 + 49 = x + 380$

2. $7 \cdot (x + 5) = 7 \cdot 38$

3. $23 + 36 + x = 24 + 36$

4. $23 \cdot (x - 36) = 46$

5. $23 \cdot 36 + x = 24 \cdot 36$

6. $12 \cdot 84 = 2 \cdot 84 + A$

7. $\dfrac{5}{16} + \dfrac{2}{3} = 2x + \dfrac{5}{16}$

8. $14Z = 7 \cdot 48$

9. $17 = 17(x - 12)$

10. $4 \cdot (x - 19) = 8$

How Is Reasoning about a Pan Balance Related to Algebraic Equation Solving?

1. For the pan balance shown below, the left side weighs the same as the right side, each small block weighs 1 ounce, and each block labeled x weighs x ounces. Explain how to determine the value of x by reasoning about the pan balance.

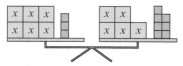

2. See if you can relate your reasoning in part 1 to steps you could take to solve the equation $6x + 3 = 5x + 7$.

3. If this balances:

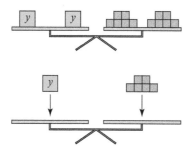

   What about this? Will it tilt to the left or to the right, or will it balance? Why?

4. If this balances:

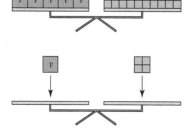

   What about this? Will it tilt to the left or to the right, or will it balance? Why?

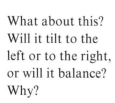

5. For the pan balance shown below, the left side weighs the same as the right side, each small block weighs 1 ounce, and each block labeled z weighs z ounces. Explain how to determine the value of z by reasoning about the pan balance.

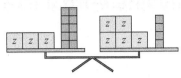

6. See if you can relate your reasoning in part 5 to steps you could take to solve the equation $3z + 10 = 5z + 4$.

7. For the pan balance shown below, the left side weighs the same as the right side, each small block weighs 1 ounce, and each block labeled w weighs w ounces. Explain how to determine the value of w by reasoning about the pan balance.

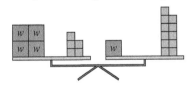

8. See if you can relate your reasoning in part 7 to steps you could take to solve the equation $4w + 5 = w + 11$.

Solving Equations Algebraically and with a Pan Balance

CCSS CCSS SMP2, 6.EE.5

1. Solve $5x + 1 = 2x + 7$ in two ways, with equations and with math drawings of a pan balance. Relate the two methods.

With equations

$$5x + 1 = 2x + 7$$

With a pan balance

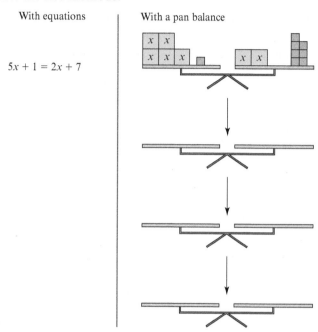

2. Solve $7 + 3x + 8 = 1 + 7x + 2$ in two ways, with equations and with math drawings of a pan balance. Relate the two methods.

3. To solve the equations in parts 1 and 2, you changed the equation, for example by subtracting the same amount from both sides and by dividing both sides by the same number. Explain why it was legitimate to change the equation in those ways and why your process led to a solution of the *original* equation.

SECTION 9.3   CLASS ACTIVITY 9-K

What Are the Solutions of These Equations?

CCSS   CCSS SMP3, 6.EE.5, 8.EE.7

Solve the next equations using the standard algebraic equation-solving process. Discuss how to interpret the outcome of that process. Think about these issues:

- Does every equation have a solution? If an equation doesn't have a solution, how can we tell that?
- Can an equation have more than one solution? Can an equation have infinitely many solutions? How could that happen?

1. $5x + 7 = 3x + 5 + x$

2. $5x + 7 = 3x + 5 + 2x$

3. $4x - 8 + x = 2x - 3$

4. $4x - 8 + x = 2x - 3 + 3x - 5$

5. Write your own equation in x that has no solutions. Explain.

6. Write your own equation in x that has infinitely many solutions. Explain.

SECTION 9.4   CLASS ACTIVITY 9-L

Solving Word Problems with Strip Diagrams and with Equations

CCSS CCSS SMP1, SMP2, 4.OA.3, 5.NF.2, 5.NF.6, 6.EE.6, 6.EE.7, 7.NS.3, 7.EE.3, 7.EE.4

The problems in this activity were inspired by problems in the mathematics textbooks used in Singapore in grades 4–6 (see [SME00] and [SMEw00], volumes 4A–6B).

1. At a store, a hat costs 3 times as much as a T-shirt. Together, the hat and T-shirt cost $35. How much does the T-shirt cost?

   Solve this problem in two ways: by using the strip diagram shown here and with algebraic equations. Explain both solution methods, and discuss how they are related.

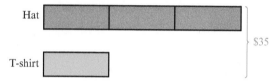

2. There are 180 blankets at a shelter. The blankets are divided into two groups. There are 30 more blankets in the first group than in the second group. How many blankets are in the second group?

   Solve this problem in two ways: by using the strip diagram shown here and with algebraic equations. Explain both solution methods, and discuss how they are related.

3. On a farm, $\frac{1}{7}$ of the sheep are gray, $\frac{2}{7}$ of the sheep are black, and the rest of the sheep are white. There are 36 white sheep. How many sheep in all are on the farm?

   Solve this problem in two ways: by using the strip diagram shown here and with algebraic equations. Explain both solution methods, and discuss how they are related.

4. Ms. Jones gave $\frac{1}{4}$ of her money to charity and $\frac{1}{2}$ of the remainder to her mother. Then Ms. Jones had $240 left. How much money did she have at first?

Solve this problem in two ways: by using the strip diagram shown here and with algebraic equations. Explain both solution methods, and discuss how they are related.

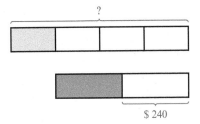

5. After Carmen spent $\frac{1}{6}$ of her money on a CD, she then had $45 left. How much money did Carmen have at first?

Solve this problem in two ways: with the aid of a strip diagram and with algebraic equations. Explain both solution methods, and discuss how they are related.

6. When a box of chocolates was full, it weighed 1.1 kilograms. After $\frac{1}{2}$ of the chocolates were eaten, the box (with the remaining chocolates) weighed 0.7 kilograms. How much did the box weigh without the chocolates?

Solve this problem in two ways: with the aid of a strip diagram and with algebraic equations. Explain both solution methods, and discuss how they are related.

7. Allie, Barbara, and Carson will divide $440 among themselves as follows: Barbara gets 2 times as much as Allie. Carson gets $\frac{1}{3}$ as much as Barbara. How much does each person get?

Solve this problem in two ways: with the aid of a strip diagram and with algebraic equations. Explain both solution methods and discuss how they are related.

8. There were 25 more girls than boys at a party. All together, 105 children were at the party. How many boys were at the party? How many girls were at the party?

   Solve this problem in two ways: with the aid of a strip diagram and with algebraic equations. Explain both solution methods, and discuss how they are related.

9. There were 10% more girls than boys at a party. All together, 168 children were at the party. How many boys were at the party? How many girls were at the party?

   Solve this problem in two ways: with the aid of a strip diagram and with algebraic equations. Explain both solution methods, and discuss how they are related.

10. A bakery sold $\frac{3}{5}$ of its muffins. The remaining muffins were divided equally among the 3 employees. Each employee got 16 muffins. How many muffins did the bakery have at first?

    Solve this problem in two ways: with the aid of a strip diagram and with algebraic equations. Explain both solution methods, and discuss how they are related.

11. Quint had 4 times as many math problems to do as Agustin. After Quint did 20 problems and Agustin did 2 problems, they each had the same number of math problems left to do. How many math problems did Quint have to do at first?

    Solve this problem in two ways: by using the strip diagram shown here and with algebraic equations. Explain both solution methods, and discuss how they are related.

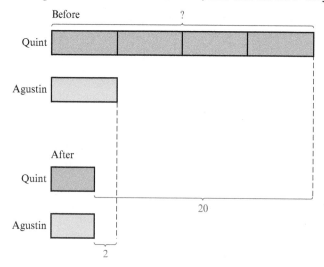

SECTION 9.4   CLASS ACTIVITY 9-M

Solving Word Problems in Multiple Ways and Modifying Problems

CCSS CCSS SMP1

1. In the morning, Ms. Wilkins put some pencils for her students in a pencil box. After a while, she found that $\frac{1}{2}$ of the pencils were gone. A little later, she found that $\frac{1}{3}$ of the pencils that were left from when she checked before were gone. Still later, Ms. Wilkins found that $\frac{1}{4}$ of the pencils that were left from the last time she checked were gone. At that point there were 15 pencils left. No pencils were ever added to the pencil box. How many pencils did Ms. Wilkins put in the pencil box in the morning?

   Solve this problem in as many different ways as you can think of, and explain each solution. Try to relate your different solution methods to each other.

2. Suppose you want to modify the pencil problem in part 1 for your students by changing the number 15 to a different number. Which numbers could you replace the 15 in the problem with and still have a sensible problem (without changing anything else in the problem)? Explain.

3. Experiment with changing some or all of the fractions—$\frac{1}{2}$, $\frac{1}{3}$, and $\frac{1}{4}$—in the pencil problem in part 1 to some other "easy" fractions. When you make a change, do you also need to change the number 15? Which changes make the problem harder? Which changes make the problem easier?

Reasoning about Repeating Patterns

CCSS CCSS SMP3, 5.OA.3

Assume that the following pattern of a square followed by 3 circles and 2 triangles continues to repeat:

1. What will be the 100th shape in the pattern? Explain how you can tell.

2. How many circles will there be among the first 100 entries of the sequence? Explain your reasoning.

3. Here are three ways that students answered part 2:

    Amanda: There are 6 circles among the first 10 shapes. Because 100 is 10 sets of 10, there will be 10 sets of 6 circles. So there are $10 \times 6 = 60$ circles among the first 100 shapes.

    Robert: My idea was like Amanda's but I got a different answer. I said there were 10 circles among the first 20 shapes. Because 100 is 5 sets of 20, there will be 5 sets of 10 circles. So there are $5 \times 10 = 50$ circles among the first 100 shapes.

    Kayla: I got the same answer as Robert, but I thought about it in a different way. The pattern repeats in sets of 6 and 3 of those 6 are circles. So half of the shapes are circles and $\frac{1}{2}$ of 100 is 50, so there are 50 circles.

    Discuss these students' ways of reasoning. Are any of their methods valid? Why or why not?

SECTION 9.5 **CLASS ACTIVITY 9-O**

Solving Problems Using Repeating Patterns

CCSS CCSS SMP1, SMP4

1. On a train, the seats are numbered as indicated below. Assume the numbering of the seats continues in this way and that all rows are arranged in the same way. Will seat number 43 be a window seat or an aisle seat? What about seat number 137? What about seat number 294? Describe different ways that you can figure out the answers to these questions.

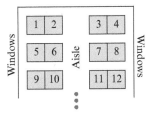

2. What day of the week will it be 100 days from today? Determine the answer with math. Explain your reasoning. How is this problem related to repeating patterns?

3. Five friends are sitting in a circle as shown. Antrice sings a song that has 22 syllables and, starting with Benton and going clockwise, points to one person for each syllable of the song. The last person that Antrice points to will be "it."

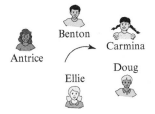

   a. Who will be "it"? Explain how to predict the answer by using math.

   b. If Fran comes and sits between Ellie and Antrice before Antrice sings her song, who will be "it"?

   c. Fran leaves, and Antrice switches to a song that has 24 syllables. Now who will be "it"? Use math to predict. What if Fran hadn't left?

4. What is the digit in the ones place of 2^{100}? Explain how you can tell.

SECTION 9.5 | **CLASS ACTIVITY 9-P** 🜍

Exploring a Growing Sequence of Figures

CCSS CCSS SMP7, 5.OA.3, 6.EE.6, 7.EE.4

In the following sequence of figures made of small circles, assume that the sequence continues by adding a green circle to the end of each of the three "arms" of a figure in order to get the next figure in the sequence.

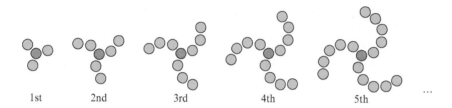

1st   2nd   3rd   4th   5th   ...

1. In the table below, write the number of small circles that the figures are made of. Imagine that the sequence of figures continues forever, so that for each counting number N, there is an Nth figure. What is an expression for the number of small circles in the Nth figure? Add this information to your table.

Position of Figure	Number of Small Circles in Figure
1	
2	
3	
4	
5	
6	
⋮	⋮
N	

2. Relate the structure of the expression you found in part (a) to the structure of the figures and use the relationship to explain why your expression makes sense.

3. How many small circles will the 38th figure in the sequence be made of? How can you tell?

4. Will there be a figure in the sequence that is made of 100 small circles? If yes, which one? If no, why not? Answer these questions in two ways: with algebraic equations and in a way that a student in elementary school who has not yet studied algebraic equations might be able to understand.

5. Will there be a figure in the sequence that is made of 125 small circles? If yes, which one? If no, why not? Answer these questions in two ways: with algebraic equations and in a way that a student in elementary school who has not yet studied algebraic equations might be able to understand.

6. Make your own growing sequence of figures and ask questions about it. Then trade your sequence and questions with a partner (or another group) and see if you can answer each other's questions.

SECTION 9.5   CLASS ACTIVITY 9-Q

How Are Expressions for Arithmetic Sequences Related to the Way Sequences Start and Grow?

CCSS   CCSS SMP7, 8.F.4

This activity will help you notice an interesting connection between the way an arithmetic sequence starts and grows and an expression for the entry in position x of the sequence. (In the next activity, you'll explain why arithmetic sequences must always have equations of a specific type.)

1. For each of the next arithmetic sequences, guess an expression for the entry in position x. Then check your guesses.

    First sequence, increasing by 4
    5, 9, 13, 17, . . . Entry in position x: _____
    Second sequence, increasing by 4
    7, 11, 15, 19, . . . Entry in position x: _____
    Third sequence, increasing by 5
    7, 12, 17, 22, . . . Entry in position x: _____

2. For each sequence in part 1, compare the expression you guessed with the way the sequence increases. What relationship do you notice?

3. For each sequence in part 1, compare the expression you guessed with the first entry of the sequence. What relationship do you notice?

4. Based on your observations in parts 2 and 3, guess the expression for the entry in position x of the next sequences. Then check your guesses.

    Fourth sequence, increasing by 3
    1, 4, 7, 10, . . . Entry in position x: _____
    Fifth sequence, decreasing by 4
    7, 3, −1, −5, . . . Entry in position x: _____

Explaining Equations for Arithmetic Sequences

CCSS CCSS SMP3, SMP7, 8.F.4

Materials Graph paper (such as Download G-6 at bit.ly/2SWWFUX) would be helpful for part 2.

1. The table below shows some entries for an arithmetic sequence whose first entry is 5 and that increases by 3.

Position	Entry
1	5
2	8
3	11
4	14
5	17
X	

a. If there were an entry in position 0, what would it be? Put it in the table.

b. Fill in the blanks to describe how to get entries in the sequence by *starting from the entry in position 0.*

- To find the entry in position 1: Start at ____ and add ____ 1 time.
- To find the entry in position 2: Start at ____ and add ____ 2 times.
- To find the entry in position 3: Start at ____ and add ____ 3 times.
- To find the entry in position 4: Start at ____ and add ____ 4 times.
- To find the entry in position 5: Start at ____ and add ____ 5 times.
- To find the entry in position X: Start at ____ and add ____ X times.

c. For each bullet in part (b), write an expression (using addition and multiplication) that corresponds to the description for finding the entry in the sequence.

Position 1 entry =

Position 2 entry =

Position 3 entry =

Position 4 entry =

Position 5 entry =

d. Let Y be the entry in position X. Use your work in parts (b) and (c) to find an equation relating X and Y.

e. Plot as many of the points (position, entry) as you can. What do you notice about the arrangement of these points?

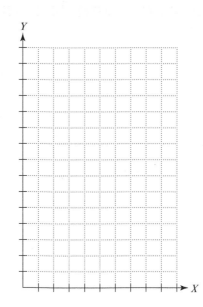

X Position	Y Entry
1	5
2	8
3	11
4	14
5	17
X	Y =

f. Discuss: How is part (b) related to going from point to point on the graph? How are the characteristics of your equation in part (d) related to your graph in part (e)?

2. Consider the arithmetic sequence whose first few entries are

$$2, 5, 8, 11, 14, \ldots$$

Let Y be the entry in position X. As in part 1, explain how to derive an equation relating X and Y. Graph some points and explain how characteristics of your equation are related to the graph and to the way the sequence grows.

SECTION 9.6 CLASS ACTIVITY 9-S 🝱

What Does the Shape of a Graph Tell Us about a Function?

CCSS CCSS SMP2, SMP4, 8.F.2, 8.F.5

1. Items (a), (b), and (c) are hypothetical descriptions of a population of fish. Each description corresponds to a population function, for which the input is time elapsed since the fish population was first measured, and the output is the population of fish at that time. Match the descriptions of these population functions to the graphs and the tables. In each case, explain why the shape of the graph fits with the description of the function and the table for the function.

   a. The population of fish rose slowly at first, and then rose more and more rapidly.

   b. The population of fish rose rapidly at first, and then rose more and more slowly.

   c. The population of fish rose at a steady rate.

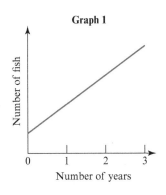
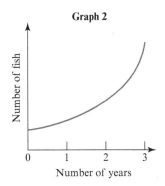
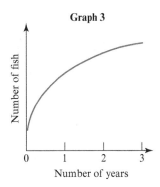

Table X	
Years	Number of Fish
0	2000
1	8000
2	10,000
3	11,000

Table Y	
Years	Number of Fish
0	2000
1	5000
2	8000
3	11,000

Table Z	
Years	Number of Fish
0	2000
1	3000
2	5000
3	11,000

2. Hot water is poured into a mug and left to cool. This situation gives rise to a temperature function for which the input is the time elapsed since pouring the water into the mug and the output is the temperature of the water at that time. The graph of this function is one of the three graphs shown next. Which graph do you think it is, and why? For each graph, describe how water would cool according to that graph.

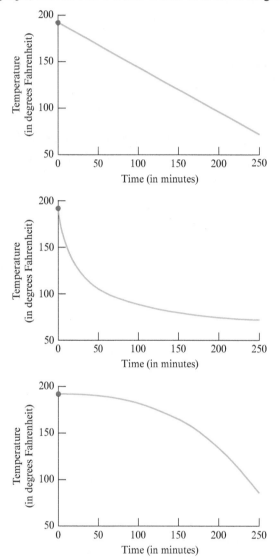

Which graph is the correct one for part 2? See page CA-238.

3. Suppose that oil escaping from an uncapped oil well below the sea floor creates a circular oil slick floating on the surface of the sea. Assume that the oil slick remains circular and that it always has the same thickness, so the diameter of the oil slick grows as more and more oil escapes from the well.

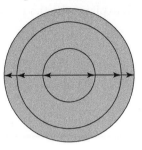

This situation gives rise to a diameter function, in which the inputs are volumes of oil in the circular oil slick and the outputs are the diameter of the oil slick when it has that volume of oil. Which graph below could be the graph of this diameter function? Explain!

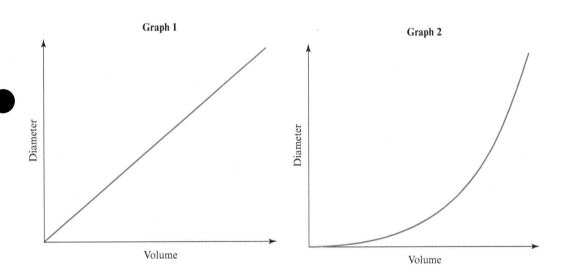

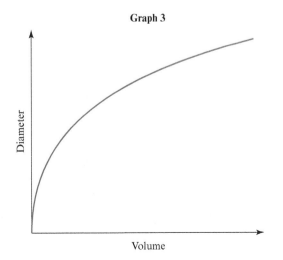

4. There are three water tanks: Tank A is in the shape of a cone, Tank B is a cylinder, and Tank C is also a cone but inverted compared to Tank A. Each tank gives rise to a height function, for which the inputs are the volume of water in the tank and the outputs are the height of the water in the tank. For each of the three height functions, sketch a graph to show the shape of the function and explain why the graph must have that shape.

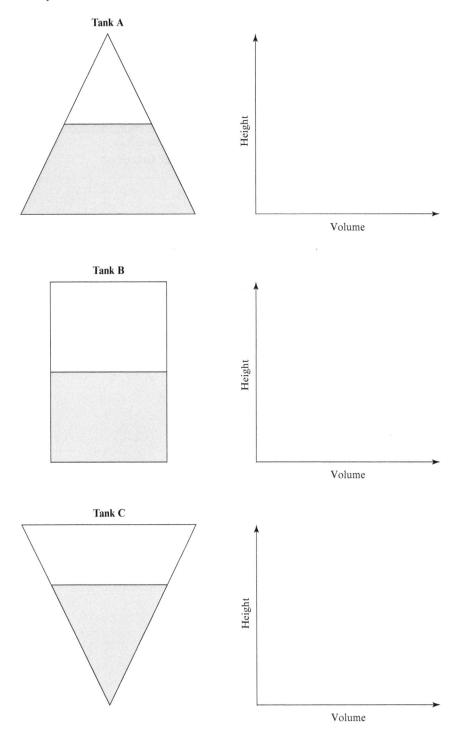

SECTION 9.6 **CLASS ACTIVITY 9-T** 🍎

Graphs and Stories

CCSS CCSS SMP3, SMP4, 8.F.2, 8.F.5

Materials Graph paper (such as Download G-6 at bit.ly/2SWWFUX) would be helpful for part 4.

1. A tagged manatee swims up a river, away from a dock. Meanwhile, the manatee's tag transmits its distance from the dock. This situation gives rise to a distance function for which the input is the time since the manatee first swam away from the dock and the output is the manatee's distance from the dock at that time. The graph of this distance function is shown below.

   Write a story about the manatee that fits with this graph. Explain how features of the graph fit with your story.

   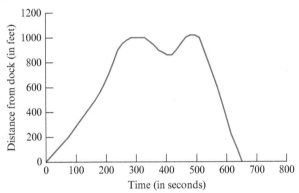

2. Carl started to drive from Providence to Boston, but after leaving he realized that he had forgotten something and drove back to Providence. Then Carl got back in his car and drove straight to Boston. This scenario gives rise to a distance function whose input is time elapsed since Carl first started to drive to Boston and whose output is Carl's distance from Providence. Could the next graph be the graph of the distance function described? Why or why not? If not, draw a different graph that could be the graph of the distance function. (Boston is 50 miles from Providence.)

   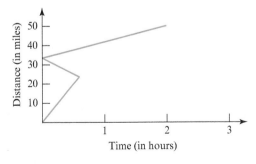
   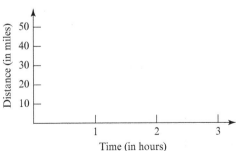

3. Here is what happened when Jenny ran a mile in 10 minutes. She got off to a good start, and ran faster and faster. Then all of a sudden, Jenny tripped. Once Jenny got back up, she started to run again, but at a slower pace. But near the end of her mile run, Jenny picked up some speed.

   The next graph is supposed to fit with the story about Jenny's mile run. What is wrong with this graph? (Look for several errors.)

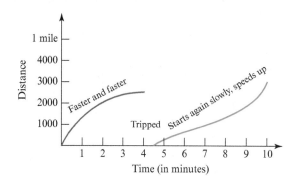

4. Sketch graphs that could be the graphs of two functions related to Jenny's mile run in part 3: a distance function and a speed function.

The correct graph on p. CA-234 is the second graph.

SECTION 9.6   CLASS ACTIVITY 9-U

How Does Braking Distance Depend on Speed?

CCSS   CCSS SMP4, SMP7, 8.F.2, 8.F.5

If you step on the brakes when you are driving, how far will you go until you come to a complete stop? The distance depends on the speed at which you are driving. Consider a *braking distance function*, which has independent variable s, the speed in miles per hour at which the car is traveling when the brakes are applied, and dependent variable d, the distance in feet which the car travels until it comes to a complete stop. (Note that we are not taking reaction time into account.)

1. Examine the graphs below and make an educated guess: Which of these graphs do you think a graph of the braking distance function will look like? Why? Which cannot possibly be the graph of the braking distance function? Why not?

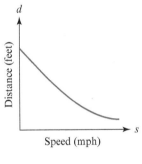

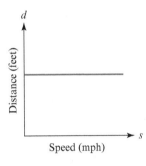

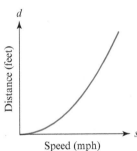

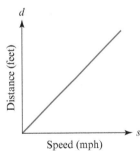

2. The braking distance function has the following property: if you double your speed, your braking distance will multiply by 4.

   a. Based on the property, which graph in part (a) represents the braking distance function? Explain how you can tell by reasoning about the graphs.

   b. Based on the property, which of the following equations could represent the braking distance function? Explain how you can tell by reasoning about the equations.

      i. $4d = 2s$

      ii. $d = 2s$

      iii. $d = \dfrac{1}{20}s^2$

SECTION 9.6   CLASS ACTIVITY 9-V

Is It a Function?

CCSS   CCSS SMP6, 8.F.1

Remember that a function is a rule that assigns one output to each allowable input.

1. Explain why each of the next proposed functions is *not* a function. Then modify the allowable inputs and rule so as to describe a function that is relevant to the given context.

   a. **Context:** Shopping at a local store.
      Proposed allowable inputs: Whole numbers between 0 and 100.
      Proposed rule: To an input number, N, associate the total cost of N items at the local store.

   b. **Context:** The number line.
      Proposed allowable inputs: Positive numbers that are less than 3.
      Proposed rule: To an input number, x, associate the number that is x units away from 5 on the number line.

   c. **Context:** Melissa drives from Indianapolis to Louisville. Her speed never exceeds 78 miles per hour.
      Proposed allowable inputs: Non-negative numbers that are less than 78.
      Proposed rule: To an input number, S, associate Melissa's distance from Indianapolis when her speed is S miles per hour.

2. For which of the following equations is y a function of x if x is allowed to vary over all real numbers? For which is x a function of y if y is allowed to vary over all real numbers?

   a. $x + 5y = 15$

   b. $x^2 + y = 100$

   c. $x^2 + y^2 = 100$

   d. $2 \cdot |x| = |y|$

3. For each example, determine if there could be a function that has those inputs and associated outputs. If so, describe a rule for such a function; if not, explain why not.

Example 1	
Input: 1	→ Output: 1
Input: 1.98	→ Output: 1
Input: 2	→ Output: 1
Input: 2.33	→ Output: 1
Input: 2.7	→ Output: 1
Input: 3.4	→ Output: 1
Input: 3.6	→ Output: 1

Example 2	
Input: 1	→ Output: 1
Input: 1.98	→ Output: 2
Input: 2	→ Output: 2
Input: 2.33	→ Output: 2
Input: 2.7	→ Output: 3
Input: 3.4	→ Output: 3
Input: 3.6	→ Output: 4

Example 3	
Input: 1	→ Output: 1
Input: 1.98	→ Output: 1
Input: 2	→ Output: 2
Input: 2.33	→ Output: 2
Input: 2.7	→ Output: 2
Input: 3.4	→ Output: 3
Input: 3.6	→ Output: 3

Example 4	
Input: 1 → Output: 1	
Input: 1 → Output: 1.98	
Input: 2 → Output: 2	
Input: 2 → Output: 2.33	
Input: 2 → Output: 2.7	
Input: 3 → Output: 3.4	
Input: 3 → Output: 3.6	

Modeling Linear Relationships with Variables and Equations

CCSS   CCSS SMP4, SMP8, 8.F.4

Remember these points when using variables and formulating equations:

- Variables stand for numbers; they are not labels. You can define a variable such as x by saying "let x be the number of . . . ."

- An equation is a statement that the quantities on the left and right of the equal sign are equal to each other. This is more specific than just saying that the quantities go together or correspond.

1. To make concrete, you need 3 times as much sand as cement, no matter how much cement you use. When Aaron was asked to define variables and describe a function about concrete, here's what he wrote:

   $S =$ sand, $C =$ cement
   $3S = C$

   Discuss Aaron's work. How could he revise it? Make a math drawing to help you explain.

2. To make a specific hue of purple paint a paint, company mixes 3 parts red paint with 2 parts blue paint, where all parts are the same number of quarts but can be any number of quarts.
   a. Define variables and describe a paint function that relates quantities of red and blue paint the company could use. Find and explain an equation for your function.

   b. Use your equation from part (a) to produce a table. Check that the values in your table fit with the information about the purple paint.

3. For each of the following, make a table to show how two quantities in the situation vary together. Then define two variables and write and explain an equation to show how the variables are related.
   a. A mail order bead company sells beads for $10 per pound. The shipping is $7 for any amount of beads.

   b. The bead company earns $4 for each pound of beads it sells. The company has weekly expenses of $500 (no matter how many beads it sells).

4. The bead company got 150 pounds of turquoise beads and figures it will sell 8 pounds of turquoise beads every day.

Examine the two different ways that students made tables to relate quantities.

Work of Student 1	
Number of Days Elapsed	Number of Pounds Left
1	$150 - 8 = 142$
2	$142 - 8 = 134$
3	$134 - 8 = 126$
4	$126 - 8 = 118$
5	$118 - 8 = 110$

Work of Student 2	
Number of Days Elapsed	Number of Pounds Left
1	$150 - 8$
2	$150 - 2 \cdot 8$
3	$150 - 3 \cdot 8$
4	$150 - 4 \cdot 8$
5	$150 - 5 \cdot 8$

Discuss the two students' work. Which will be more useful for formulating an equation in two variables? Why?

5. Another bead company sells beads for $12 for every 5 pounds of beads. The shipping is $4 for any amount of beads.

Make a table to show how two quantities in this situation vary together. Then define two variables and write and explain an equation to show how the variables are related.

SECTION 9.7   CLASS ACTIVITY 9-X

Interpreting Equations for Linear Relationships

CCSS   CCSS SMP2, SMP4, 8.F.4

1. At the yogurt store, the cost of a cone of frozen yogurt depends on how much yogurt is in the cone. The cost and the amount of yogurt are related by the equation

$$C = 0.55Y + 1.15$$

where C is the cost in dollars and Y is the number of ounces of yogurt in the cone. Explain how to interpret 0.55 and 1.15 in terms of the situation.

2. A phone company charges $0.50 for a 5-minute phone call plus an additional $0.06 for each additional minute after that. Some students are trying to write equations to describe the relationship between the cost and the number of minutes of a phone call.

    Discuss the following students' ideas and questions. Include clarifications or modifications as part of your discussion.

    a. Niles writes the equation $C = 0.06(T - 5) + 0.50$.

    b. Chad asks what the 0.06 multiplied by $T - 5$ stands for.

    c. Francine wants to know what happens if T is 4. Does the equation work?

    d. Anja writes the equation $C = 0.06T + 0.50$.

    e. Quowanna wants to know how Niles and Anja defined their variables. She wonders if they defined T differently.

3. A grocer will spend $150 on beans and tomatoes combined. Beans cost $2 per pound and tomatoes cost $3 per pound. Let B be a number of pounds of beans and T the corresponding number of pounds of tomatoes the grocer could buy.

Discuss and elaborate on the following students' ideas.

a. Mariah says she has a simple equation but it's not in the form $B =$ (some expression) or $T =$ (some expression).

b. Chris tries to formulate an equation in the form

$$T = 50 - (\text{something}) \cdot B$$

He found the 50 by reasoning that if the grocer buys no beans, he can buy 50 pounds of tomatoes.

c. DeShun says that for every additional 3 pounds of beans, the grocer must buy 2 fewer pounds of tomatoes, and she thinks this can help Chris find his equation.

4. Initially, a tank had 80 gallons of water in it when water started flowing out of it. Let W be the number of gallons of water left in the tank T seconds after water started flowing out of the tank. Suppose that W and T are related by the equation

$$W = 80 - T$$

Some students objected to this equation, saying that the 80 is gallons and the T is seconds and it doesn't make sense to subtract seconds from gallons. Discuss!

SECTION 9.7   CLASS ACTIVITY 9-Y

Are There Different Kinds of Decreasing Relationships?

CCSS   CCSS SMP4, 8.F.3

Materials Graph paper (such as Download G-6 at bit.ly/2SWWFUX) would be helpful for part 2.

Consider these two relationships:

Relationship 1: There are a number of hoses, all of the same size. Water flows out of all of these hoses at the same constant rate. Using 4 hoses, it takes 6 hours to fill an empty tub with water. Let x be the number of hoses used to fill the empty tub and let y be the number of hours it takes to fill the tub using that many hoses.

Relationship 2: Initially, there are 100 liters of water in a tub. Then water starts to flow out of the tub at a constant rate. After 20 minutes the tub is empty. Let x be the number of minutes since water starts flowing out of the tub and let y be the number of liters of water in the tub at that time.

1. In both relationship 1 and relationship 2, as x increases, what happens to y?

2. Make a table, draw a graph, and write an equation for Relationship 1 and for Relationship 2. What kinds of relationships are they and how can you tell?

How Do Patterns in Tables Reflect Different Kinds of Relationships?

CCSS   CCSS SMP4, 8.F.3, 8.F.4

1. A group is throwing a benefit concert to raise money for a charity. Let x be the number of tickets the group might sell and let y be the net amount of money the group will raise if they sell that many tickets.

x	y
0	-500
20	-100
50	500
100	1500
200	3500

a. Could the relationship be linear or not? How can you tell?

b. Discuss the entries in the table that have a negative y-coordinate. Interpret these entries in terms of the scenario.

c. How much are tickets being sold for? How can you tell?

d. Where should the graph cross the x-axis and what is the significance of this?

2. A marble is dropped from the top of a tall building. Let x be the number of seconds elapsed since the marble was dropped and let y be the number of feet that the marble is above the ground at that time.

x	y
0	256
1	240
2	192
3	112
4	0

   a. Could the relationship be linear or not? How can you tell?

   b. How tall is the building? When does the marble hit the ground?

   c. Describe how the marble falls.

   d. What do you notice about how y changes as x increases by 1? Do you notice any pattern?

   e. In part (d) you found the changes in the y values in the table. Now find the changes in the changes in the y values in the table. What do you notice?

3. A bank account was opened, and some money was put into it. Let x be the number of years since the money was put into the account, and let y be the amount of money in the account.

x	y
0	100
5	200
10	400
15	800
20	1600

   a. Could the relationship be linear or not? How can you tell?

   b. How much money was put into the account initially?

   c. Describe how the amount of money grows.

SECTION 9.7 CLASS ACTIVITY 9-AA

What Kind of Relationship Is It?

CCSS CCSS SMP4

For each of the tables below, determine what kind of relationship the table exhibits and explain how you can tell. You do not need to find equations for the relationships.

Table A	
x	y
1	3
2	8
3	15
4	24
5	35

Table B	
x	y
1	7
3	10
7	16
13	25
21	37

Table C	
x	y
1	3
2	6
4	12
8	24
16	48

Table D	
x	y
1	3
2	6
3	12
4	24
5	48

Table E	
x	y
1	25
2	20
3	15
4	10
5	5

Table F	
x	y
1	60
2	30
3	20
4	15
5	12

Doing Rocket Science by Reasoning about the Structure of Quadratic Equations

CCSS CCSS SMP7

1. Rocket scientists determined that t seconds after launch, a rocket's height, h, in feet above the ground will be given by the equation

$$h = 16(2 + t)(22 - t)$$

   a. Describe the structure of the expression on the right of the equal sign.

   b. Reason about the structure of $16(2 + t)(22 - t)$ to determine which values for t will make the expression have a value of 0. To help your thinking: when you multiply numbers, how can the result be 0?

   c. Why would the rocket scientists want to know when the expression has a value of 0?

2. The rocket scientists determined that another equation for the height of the rocket is

$$h = 2304 - 16(t - 10)^2$$

a. Describe the structure of the expression on the right side of the equal sign.

b. Reason about the structure of $2304 - 16(t - 10)^2$ to determine the largest value it can have and to determine the value of t at which this occurs.

To help your thinking, consider these questions:

- Why can $16(t - 10)^2$ never be negative?
- How can you use the structure of $16(t - 10)^2$ to determine for which value of t it is 0?

c. Why would the rocket scientists want to know the largest value of the expression?

Reasoning about the Structure of Quadratic Equations to Solve Profit Problems

CCSS CCSS SMP7

A company that sells Gizmos knows that its annual profit depends on the price it sells its Gizmos for. As the company changes the price of Gizmos, its profit changes.

The company has found three ways to write an equation for a profit function (by writing equivalent expressions). In each case, the independent variable, x, is the price of a Gizmo in cents and the dependent variable, y, is the company's annual profit in thousands of dollars.

1. Explain how to reason about the structure of Equation 1 to tell you when the company's profit is 0.

$$y = 2(x - 10)(80 - x) \tag{1}$$

2. Explain how to reason about the structure of Equation 2 to tell you how the company can make its maximum profit.

$$y = 2450 - 2(x - 45)^2 \tag{2}$$

3. Explain how to use Equation 3 to tell you what would happen if the company gave Gizmos away for free.

$$y = -2x^2 + 180x - 1600 \tag{3}$$

4. What do parts 1, 2, and 3 tell you about the coordinates of points A, B, C, and D in the graph of the profit function?

Graph of the Gizmo Company's
profit function

$y = 2(x - 10)(80 - x)$

$y = 2450 - 2(x - 45)^2$

$y = -2x^2 + 180x - 1600$

Annual profit
(thousands
of dollars)

Price of a
Gizmo (cents)

SECTION 9.8 CLASS ACTIVITY 9-DD

Sums of Powers of Two

Is there a quick way to add a bunch of consecutive powers of 2? This activity will help you find and explain a formula.

1. Calculate the next sums.

$$1 + 2 = \underline{\hspace{1cm}}$$
$$1 + 2 + 4 = \underline{\hspace{1cm}}$$
$$1 + 2 + 4 + 8 = \underline{\hspace{1cm}}$$
$$1 + 2 + 4 + 8 + 16 = \underline{\hspace{1cm}}$$

2. Based on part 1, predict the sum of the following geometric series without adding all the terms:

$$1 + 2 + 4 + 8 + 16 + 32 + 64 + 128 + 256 + 512$$

3. Based on part 1, predict an expression in terms of N for the following geometric series (fill in the blank with an appropriate expression):

$$1 + 2 + 2^2 + 2^3 + 2^4 + \ldots + 2^N = \underline{\hspace{1cm}}$$

4. Use the figure below to help you explain why your expression in part 3 should be true.

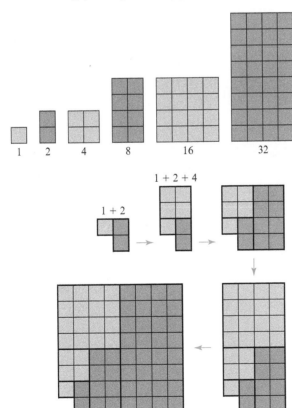

5. Here is a systematic way to find an expression for the sum of the geometric series in part 3: Let S be this sum, so that

$$S = 1 + 2 + 2^2 + 2^3 + 2^4 + \cdots + 2^N$$

Use the distributive property to write $2S$ as a series (fill in the blank with a series):

$$2S = 2 \cdot (1 + 2 + 2^2 + 2^3 + 2^4 + \cdots + 2^N)$$
$$= \underline{\hspace{5cm}}$$

Now calculate $2S - S$ in the following two ways, in terms of S and as a series:

$2S - S$ in terms of S:

$$2S$$
$$\underline{-\ S}$$

$2S - S$ as a series:

$$\underline{-1 - 2 - 2^2 - 2^3 - 2^4 + \cdots - 2^N}$$

The two results you get must be equal, so you get an equation. What does this equation tell you about S?

SECTION 9.8   CLASS ACTIVITY 9-EE

An Infinite Geometric Series

Is it possible to calculate an infinite sum of numbers? Surprisingly, the answer is yes, as you will see in this activity.

Assume that the next sequence of partially shaded squares continues indefinitely, in such a way that we get the next square in the sequence by shading $\frac{1}{2}$ of the unshaded part of a square.

1. Explain why the shaded portions of the 2nd, 3rd, and 4th squares in the sequence are

$$\frac{1}{2} + \left(\frac{1}{2}\right)^2$$

$$\frac{1}{2} + \left(\frac{1}{2}\right)^2 + \left(\frac{1}{2}\right)^3$$

$$\frac{1}{2} + \left(\frac{1}{2}\right)^2 + \left(\frac{1}{2}\right)^3 + \left(\frac{1}{2}\right)^4$$

2. What fraction of the square is shaded in the 2nd, 3rd, and 4th figures? Give each answer in simplest form. Based on your results, predict what fraction of the Nth square is shaded.

3. Based on your work in part 2, when would you reach a square that is at least 99.9% shaded?

4. Based on the sequence of shaded squares, what would you expect the infinite sum

$$\frac{1}{2} + \left(\frac{1}{2}\right)^2 + \left(\frac{1}{2}\right)^3 + \left(\frac{1}{2}\right)^4 + \left(\frac{1}{2}\right)^5 + \cdots$$

to be equal to?

5. Here is a way to calculate the infinite sum in part 4. Let S stand for this sum. In other words,

$$S = \frac{1}{2} + \left(\frac{1}{2}\right)^2 + \left(\frac{1}{2}\right)^3 + \left(\frac{1}{2}\right)^4 + \left(\frac{1}{2}\right)^5 + \cdots$$

Assuming that there is an "infinite distributive property," write $\frac{1}{2}S$ as an infinite sum (fill in the blank):

$$\frac{1}{2}S = \frac{1}{2}\cdot\left(\frac{1}{2} + \left(\frac{1}{2}\right)^2 + \left(\frac{1}{2}\right)^3 + \left(\frac{1}{2}\right)^4 + \cdots\right)$$

$$= \underline{\hspace{4cm}}$$

Now calculate $S - \frac{1}{2}S$ in two ways, in terms of S and as a series:

In terms of S: As a series:

$$S$$ $$\frac{1}{2} + \left(\frac{1}{2}\right)^2 + \left(\frac{1}{2}\right)^3 + \left(\frac{1}{2}\right)^4 + \left(\frac{1}{2}\right)^5 + \cdots$$

$$-\frac{1}{2}S$$ $$\underline{\hspace{5cm}}$$

The two results you get must be equal, so you get an equation. Solve this equation for S.

SECTION 9.8 CLASS ACTIVITY 9-FF

Making Payments into an Account

Materials You will need a calculator for this activity.

Suppose that at the beginning of every month, you make a payment of $200 into an account that earns 1% interest per month. (That is, the value of the account at the end of the month is 1% higher than it was at the beginning of the month.)

1. Make a guess: After making your 12th payment at the beginning of the 12th month, how much money will be in the account? (The remaining parts of this activity will help you calculate this amount exactly.)

2. Explain why the entries shown in the right-hand column in the table below are correct. Then fill in the remaining columns with a series similar to those in the first 3 rows. (You may use " ... " within your series in the last row.)

Month	Amount at Beginning of Month (After Payment)
1	200
2	$200 + (1.01)200$
3	$200 + (1.01)200 + (1.01)^2 200$
4	
5	
6	
⋮	⋮
12	

3. Let S be the series you wrote in the last row of the right column in part 2. Use the distributive property to write $(1.01)S$ as a series.

4. Calculate $(1.01)S - S$ in two ways: in terms of S and as a series.

5. The two results you get in part 4 must be equal, so you get an equation. Solve this equation for S. Compare the result with your guess in part 1.

SECTION 10.1   CLASS ACTIVITY 10-A

What Are Angles and Why Do We Have Them?

Discuss the following questions with a partner or small group and gather some initial thoughts on how to answer them.

1. Why do we have the concept of angle? What do we use angles for?

2. What are angles? Gather some initial ideas on how you might explain the concept of angle.

3. The angles below are the same size, but some students will think the angle on the right is bigger. Why might students think that? What ideas do you have for helping students understand that these angles are the same size?

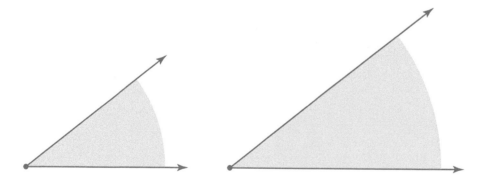

Measuring Angles with Folded Circle Protractors

CCSS CCSS SMP3, SMP5, 4.MD.5, 4.MD.6, 4.MD.7

Materials Each person will need at least six circles cut from a piece of paper or from Download 10-1 at bit.ly/2SWWFUX

1. Fold circles to create the following angles. Crease your folds to make the angles clearly visible. In each case, explain why your folding creates angles of that size.
   a. 180°

   b. 90°

   c. 45°

   d. 60°

   e. Fold at least two more circles to create some other angles!

2. Now think of each folded circle as a protractor. Use folded-circle-protractors to determine the measures of angles *a*, *b*, *c*, *d*, *e*, and *f* below.

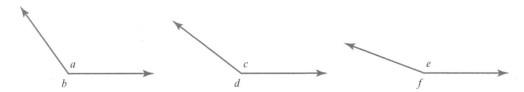

3. Explain how to use your folded circle protractors from part 1 to make the following angles.

   a. 315°

   b. 165°

   c. 105°

SECTION 10.1 **CLASS ACTIVITY 10-C** 🍎

Angles Formed by Two Lines

CCSS CCSS SMP7, 7.G.5

1. The figure below shows three pairs of lines meeting (or you may wish to think of this as showing one pair of lines in three situations, when the lines are moved to different positions). In each case, how do opposite angles *a* and *c* appear to be related, and how do opposite angles *b* and *d* appear to be related?

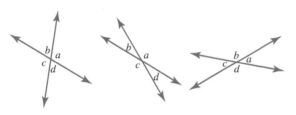

2. Do you think the same phenomenon you observed about opposite angles in part 1 will hold for *any* pair of lines meeting at a point? Do you have a convincing reason why or why not?

3. Explain *why* what you observed in part 1 about opposite angles must always be true by using the fact that an angle formed by a straight line is 180°. What does this fact tell you about several pairs of angles?

Angles Formed When a Line Crosses Two Parallel Lines

CCSS   CCSS SMP3, 8.G.5

A pair of lines marked with arrows indicates that the lines are parallel.

Given two parallel lines and a transversal line that crosses the two parallel lines, as shown below, the Parallel Postulate says that $a = a'$, $b = b'$, $c = c'$, and $d = d'$.

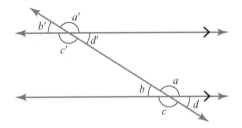

1. Given that lines m and n are parallel, explain how to use the Parallel Postulate and what we know about opposite angles to relate angles e and f.

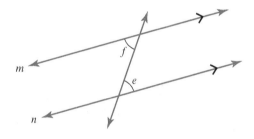

2. Given that lines p and q are parallel, explain how to use the Parallel Postulate to relate angles g and h.

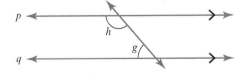

SECTION 10.1 **CLASS ACTIVITY 10-E**

How Are the Angles in a Triangle Related?

CCSS CCSS SMP3, 8.G.5

Materials Each small group will need a ruler and scissors for this activity.

Work with a small group. Each person in your group should do the following:

1. Using a ruler, draw a large triangle that looks different from the triangles of other group members. Cut out your triangle. Label the three angles a, b, and c.

2. Tear or cut all three corners off your triangle. Then put the angles together vertex to vertex, without overlaps or gaps. What do you notice? What does this show you about the angles in the triangle?

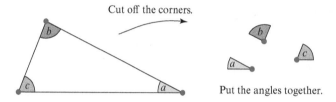

3. Discuss why the "putting the angles together" method of part 2 is not a proof. Consider these points:

   • Does the method show *exactly* how the angles are related?

   • Has every triangle been considered?

SECTION 10.1 **CLASS ACTIVITY 10-F**

Drawing a Parallel Line to Prove That the Angles in a Triangle Add to 180°

CCSS CCSS SMP3, 8.G.5

This activity will show you a way to prove that the angles in a triangle add to 180°.

1. Given any triangle with (corner) points A, B, and C, let a, b, and c be the angles in the triangle at A, B, and C, respectively. Consider the line through A parallel to the side BC that is opposite A.

   What can you say about the three adjacent angles at A that are formed by the triangle and the line through A?

   What can you conclude about the sum of the angles in the triangle?

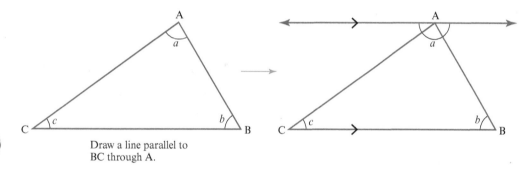

Draw a line parallel to BC through A.

2. What if you used a different triangle in part 1? Would you still reach the same conclusion?

Walking and Turning to Explain Relationships among Exterior and Interior Angles of Triangles

CCSS   CCSS SMP3, 8.G.5

Materials You will need sticky notes and, if available, masking tape.

This activity will show you a way to understand why the angles in a triangle add to 180°. It is best done as a demonstration for the whole class.

1. Put three "dots" (sticky notes) labeled A, B, and C on the floor to create a triangle. If possible, connect them with masking tape. Label a point P on the line segment between A and C.

2. Choose two people: one to be a *walker* and one to be a *turner*. The rest are *observers*.

   The walker's job: Stand at point P, facing point A. Walk all the way around the triangle, returning to point P.

   The turner's job: Stand at one fixed spot, and face the same direction that the walker faces at all times. This means that when the walker turns at a corner, you should turn in the same way.

   The observers' job: Observe the walker and the turner, and make sure that they really are facing the same direction at all times.

   Repeat the walking and turning described above until everyone can confidently answer the following questions:

   a. Let's say that the walker and turner were facing north when the walker began walking around the triangle. Which directions did the turner face during the experiment? Were any directions left out? Were any directions repeated?

   b. What was the full angle of rotation of the turner when the walker walked once all the way around the triangle, returning to point P?

3. On the drawing below, show which angles the walker turned through at the corners of the triangle. Label these angles *d*, *e*, and *f*.

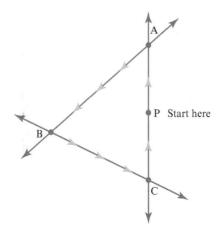

4. Based on your answer to part 2(b), what can you say about the value of $d + e + f$?

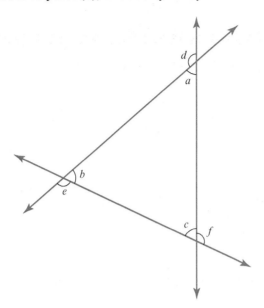

5. Check your answer to the previous part by using a protractor to find the angles d, e, f and adding them.

What if you used a different triangle? The values of $d, e,$ and f might be different, but what about $d + e + f$?

6. You should have just found that the sum of the exterior angles of a triangle is 360°. In other words, $d + e + f = 360°$, where $d, e,$ and f are the exterior angles, as shown in the figure above.

Use the formula $d + e + f = 360°$ to explain why the sum of the interior angles in a triangle is equal to 180°. In other words, show that $a + b + c = 180°$, where $a, b,$ and c are the interior angles of a triangle, as shown in the figure.

Hint: What do you notice about $a + d, b + e,$ and $c + f$? Can you use this somehow?

7. What if you used a different triangle in this activity? Would you still reach the same conclusion?

SECTION 10.1 **CLASS ACTIVITY 10-H** 🍎

Angle Problems

CCSS CCSS SMP1

In some (but not all) of these problems, it will be helpful to add or to extend one or more lines or line segments.

1. Determine a formula for angle x in terms of angles a and b. Explain why your formula is valid (without measuring any angles).

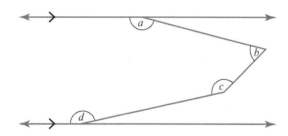

2. Given that the lines marked with arrows are parallel, determine the sum of the angles $a + b + c + d$. Explain why your answer is valid (without measuring any angles). See if you can find more than one explanation!

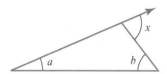

3. Determine the sum of the angles, $a + b + c + d + e$. Explain why your answer is valid (without measuring any angles). See if you can find two explanations!

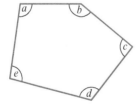

4. In part 3, what if the shape had N angles instead of 5 angles? Find a formula for the sum of the angles in terms of N and explain your reasoning.

SECTION 10.1 CLASS ACTIVITY 10-I

Students' Ideas and Questions about Angles

CCSS CCSS SMP3

Discuss the following ideas and questions that students had about angles.

1. Harry has learned that the angles in a triangle add to 180° and that there are 360° in a circle. Harry wonders about the drawing below: "Shouldn't the circle be less than 180° because it is inside the triangle?"

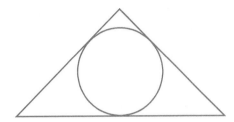

2. Sam says that a circle has infinitely many degrees because you can think of a degree as being like a little wedge. By making the wedges smaller, you can fit more and more wedges in the circle.

Kaia counters that you can't have an angle smaller than 1°.

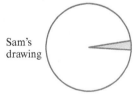

Sam's drawing

Eratosthenes's Method for Determining the Circumference of the Earth

CCSS   CCSS SMP4

The figure below shows a cross-section of the earth. At noon on June 21, the sun is directly overhead at location A, so that the sun's rays are perfectly vertical there. At the same time, 500 miles away at location B, the sun's rays make a 7.2° angle with the tip of a vertical pole (shown not to scale), which was determined by considering the shadow that the pole casts. Because the sun is far away, sun rays at the earth are (approximately) parallel. Use this information to determine the circumference of the earth, explaining your reasoning.

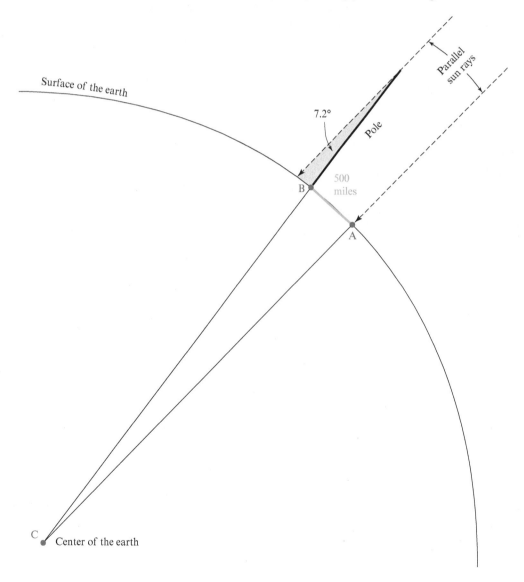

SECTION 10.2 **CLASS ACTIVITY 10-K**

Why Do Spoons Reflect Upside Down?

CCSS CCSS SMP4

Materials A large, reflective spoon would be helpful for this activity.

When you look at your reflection in the bowl of a spoon, you will notice that (in addition to looking quite distorted) your image will appear upside down. Have you ever wondered why? We will explore this now.

The figure below shows a person looking into the bowl of a (very large) spoon. Only a cross-section of the spoon is shown. The lines shown are the *normal lines* to the spoon at the points A, B, and C. Use these normal lines and the laws of reflection to determine what the person sees when she looks at points A, B, and C. (Assume that the person sees light that enters the center of her eye.)

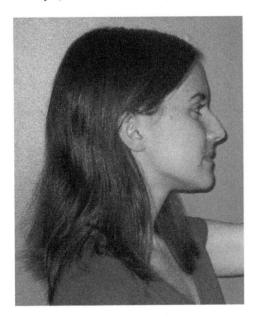
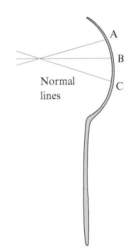

Normal
lines

How Big Is the Reflection of Your Face in a Mirror?

CCSS   CCSS SMP4

Materials You will need a mirror and a ruler for the first part of this activity.

How big is the reflection of your face in a mirror? Is it the same size as your face, or is it larger or smaller, and if so, how much? The answers to these questions may surprise you.

1. Use a ruler to measure the length of your face from the top of your forehead to your chin. Now hold a mirror *parallel* to your face and measure the length of your face's reflection in the mirror, from the top of the forehead to the chin. Measure carefully!

   Hold the mirror closer to your face or farther away (but always *parallel* to your face), and repeat the measuring processes just described. The *position* of your reflection will probably change, but does the *size* of your reflection change or not? The answer may surprise you.

   Compare the length of your face and the length of your reflected face in the mirror. How do these lengths appear to be related?

2. The figures on the next page show a side view of a person looking into a mirror. Using the laws of reflection, determine what the person will see at each of the points A, B, and C. We see objects by seeing the light that travels from the object to our eyes. So a person looking at a particular point sees the light that travels in a straight line from that point to the person's eye. To determine what a person sees at a point in a mirror, you must determine where the light at that point came from. For this you will need the laws of reflection.

3. Use the laws of reflection to show where the person looking into the mirror will see the top of her forehead and where she will see the bottom of her chin. Measure the length of the person's face and the length of her reflected face in the mirror. How do these lengths appear to be related? Your result should fit with what you discovered in part 1.

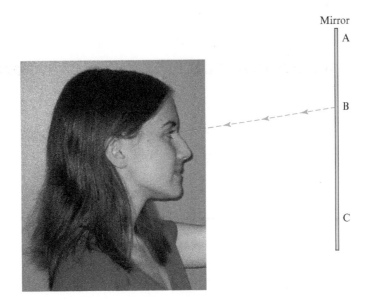

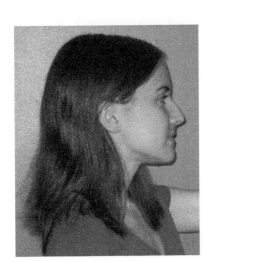

SECTION 10.3 CLASS ACTIVITY 10-M

Points That Are a Fixed Distance from a Given Point

CCSS CCSS SMP2

Materials You will need a ruler for this activity.

1. Use a ruler to draw five different points that are 1 inch away from the point P:

• P

Now draw five more points that are 1 inch away from the point P. If you could keep drawing more and more points that are 1 inch away from point P, what shape would this collection of points begin to look like?

2. Ask a person to point to a particular point in space. Call that point P. Using a ruler, find several other locations in space that are 1 ruler-length away from point P. (A ruler-length might be 12 inches or 6 inches, depending on your ruler.) Try to visualize all the points in space that are 1 ruler-length away from your point P. What shape do you see?

SECTION 10.3 **CLASS ACTIVITY 10-N**

Using the Mathematical Definition of Circles

CCSS CCSS SMP4

Materials You will need a compass for this activity. Some string would also be useful.

1. Use the definition of a circle to explain why a compass can be used to draw a circle. How is the radius of the circle related to the compass?

2. Explain how to use a piece of string and a pencil to draw a circle. Use the definition of a circle to explain why your method can be used to draw a circle.

3. Suppose that, 1 hour and 45 minutes ago, a prisoner escaped from a prison located at point P, shown on the map below. Due to the terrain and the fact that no vehicles have left the area, police estimate that the prisoner can go no faster than 4 miles per hour. Show all places on the map where the prisoner might be at this moment. Explain.

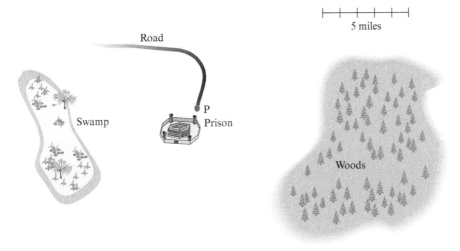

4. Some treasure is described as buried under a spot that is 30 feet from a spot marked X and 50 feet from a spot marked O. Use this information to help you show where the treasure might be buried on the map below. Is the information enough to tell you *exactly* where the treasure is buried? Explain.

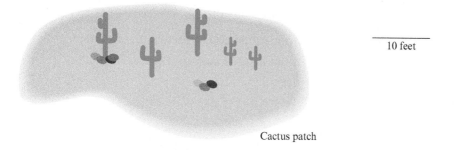

10 feet

Cactus patch

X   O

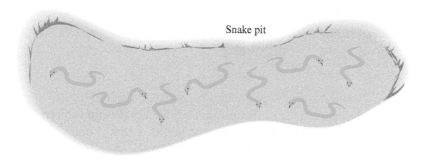

Snake pit

SECTION 10.3 CLASS ACTIVITY 10-O

The Global Positioning System (GPS)

CCSS CCSS SMP4

Materials You will need string for this activity.

The activity is best as a whole-class demonstration.

GPS units determine their location by receiving information from satellites that orbit the earth. This activity simulates the process.

Suppose a GPS unit learns that it is a certain distance from satellite 1 and another certain distance from satellite 2.

1. Choose two people to represent satellites 1 and 2, and choose a third person who will show all possible locations of the GPS unit. The GPS person should stand between satellites 1 and 2.

2. Cut two pieces of string, representing the distances from satellites 1 and 2 to the GPS unit.

3. Satellites 1 and 2 should each hold one end of their piece of string, and the GPS person should hold the other ends of the two pieces of string in one hand, pulling the strings tight. (Everyone may have to adjust positions so that it is possible to do this. Once suitable positions are found, everyone should stay fixed in their positions.)

4. The designated GPS person will now be able to show all possible locations of the GPS unit by moving the strings, while keeping them pulled tight and held in one hand (and while satellites 1 and 2 remain fixed in their positions).

5. Describe the shape of all possible locations of the GPS unit. How is this related to the intersection of two spheres? Are two satellites enough to determine a location?

6. Now suppose that there is also a third satellite beaming information to the GPS unit. Choose a person to represent satellite 3, and cut a piece of string to represent the distance of satellite 3 to the GPS unit.

7. Satellite 3 should hold one end of their string while the GPS person holds the other end in the same hand with the strings from satellites 1 and 2. By pulling all three strings tight, the GPS person can show all possible locations of the GPS unit. In general, there will be two such locations.

    As you've seen, with information from three satellites, a GPS unit can narrow its location to one of two points. If one of those two locations can be recognized as being in outer space, and not on the surface of the earth, then the GPS unit can report its location on the earth. This simulates the idea behind the GPS system.

SECTION 10.3 CLASS ACTIVITY 10-P

Circle Designs

CCSS CCSS SMP5, SMP7

Materials You will need a paperclip or a compass. For part 3, you might want graph paper, or use a jumbo paperclip and Download 10-2 at bit.ly/2SWWFUX

1. Although the accompanying design looks complex, it is surprisingly easy to make. Use a compass (or a paper clip) to draw a design like the one below. Say briefly how to draw the design.

2. The next design was made by drawing part of the design of part 1. Use a compass (or a paper clip) to draw a design like this one. (Color it later if you like.)

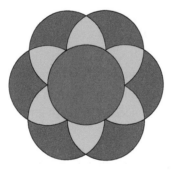

3. To draw the designs below, use the Download or draw square grids on graph paper first. Then use a compass or a jumbo paperclip. (Color the designs later if you like.)

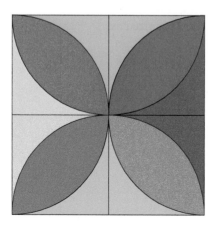 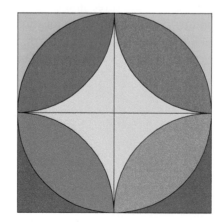

SECTION 10.4 CLASS ACTIVITY 10-Q

What Shape Is It?

CCSS CCSS SMP6, K.G.2

Is it always obvious what type a shape is? Examine and discuss the examples below.

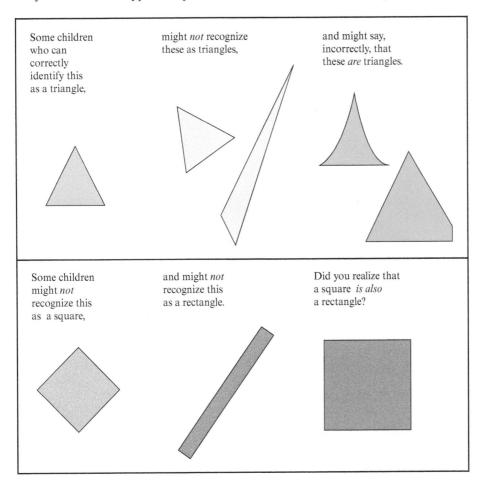

Now ponder this question: What would be a more reliable way to determine if a shape is or isn't of a specific type than simply observing what the shape looks like overall?

SECTION 10.4 · CLASS ACTIVITY 10-R

What Properties Do These Shapes Have?

CCSS CCSS SMP7, 1.G.1

1. What properties or attributes do the shapes below have (or appear to have)? What do you notice about their sides? What do you notice about their angles? What other properties or attributes do you notice? List as many as you can find!

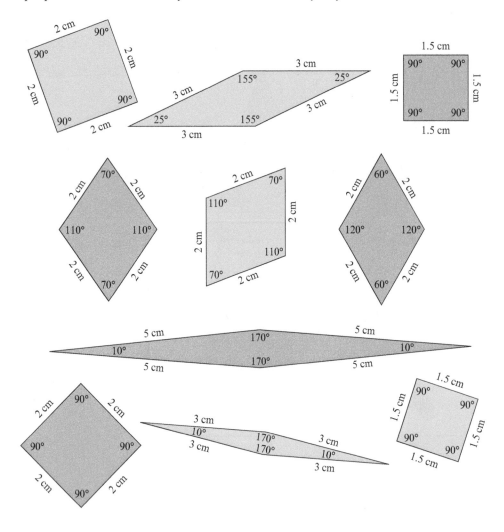

2. Which of the properties or attributes that you noticed in part 1 might be useful for deciding what type of shape they are? Which properties or aspects are *not* relevant for deciding what type of shape they are?

SECTION 10.4 CLASS ACTIVITY 10-S 🏛

How Can We Classify Shapes into Categories Based on Their Properties?

CCSS CCSS SMP6, 2.G.1, 3.G.1, 4.G.2, 5.G.3, 5.G.4

1. For each category on the next page, identify *all* those shapes on this page that have *all* the attributes listed. Note that some shapes fit into more than one category and some shapes may not fit in any category.

   Assume that sides that appear to be the same length are (but measure if you aren't sure). Assume that angles that appear to be right angles are. Assume that sides that appear to be parallel are.

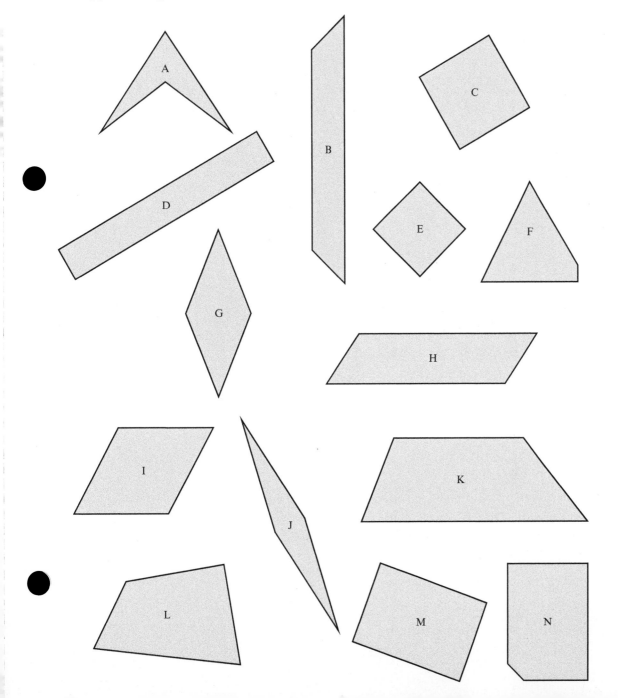

Category 1: Four sides, four right angles, all four sides the same length

Category 2: Four sides, four right angles	**Category 3:** Four sides, four right angles, opposite sides parallel	**Category 4:** Four sides, four right angles, opposite sides same length

Category 5: Four sides, all the same length	**Category 6:** Four sides, all the same length, opposite sides parallel	**Category 7:** Four sides, all the same length, opposite angles same size

Category 8: Four sides, opposite sides parallel	**Category 9:** Four sides, opposite sides same length
Category 10: Four sides, *at least* one pair parallel	**Category 11:** Four sides, *exactly* one pair parallel

Category 12: Four sides

2. Which categories in part 1 have exactly the same shapes? Try to draw a shape (using a ruler and protractor or geometry software) that will fit in one of those categories but not the other. Do you think it is possible?

    For each category, provide its common name if you know it.

3. Which categories are a subcategory of another category? How can you tell that from lists of properties? Describe or draw a diagram to show how the categories in part 1 are related to each other.

CLASS ACTIVITY 10-T

How Can We Classify Triangles Based on Their Properties?

CCSS CCSS SMP5, SMP6, 4.G.2, 5.G.3, 5.G.4

1. For each category of triangles below, find all the triangles on the next page that have the given attribute.

Category 1: All three sides same length (equilateral triangles)	**Category 2:** All three angles same size

Category 3: At least two sides same length (isosceles triangles)	**Category 4:** At least two angles same size

Category 5:
All angles are smaller than a right angle
(acute triangles)

Category 6:
Has a right angle
(right triangles)

Category 7:
Has an angle greater than a right angle
(obtuse triangles)

2. Which categories in part 1 have exactly the same triangles? Try to draw (using a ruler and protractor or geometry software) a triangle that will fit in one of those categories but not the other. Do you think it is possible?

3. How are categories 1 and 3 related? Why?

4. How are categories 5, 6, and 7 related and how are these three categories related to the category of all triangles? Why?

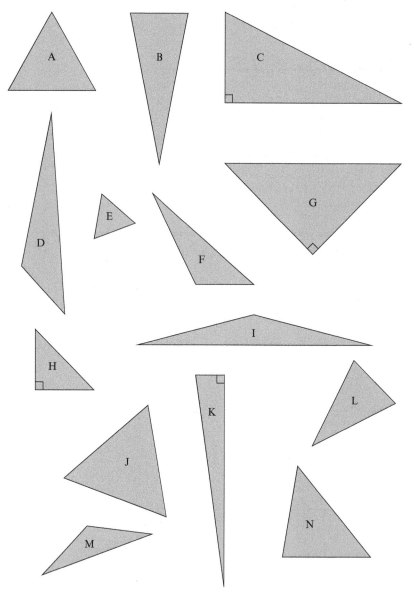

Note: Assume that sides that appear to be the same length are the same length.

SECTION 10.4 CLASS ACTIVITY 10-U

Using Venn Diagrams to Relate Categories of Quadrilaterals

CCSS CCSS SMP6, SMP7, 5.G.4

1. Imagine that you have a large collection of plastic squares and rectangles in different sizes. You might sort those squares and rectangles in different ways.

   Which of the three ways of sorting the squares and rectangles shown below fits with the *definitions* of squares and rectangles? Explain!

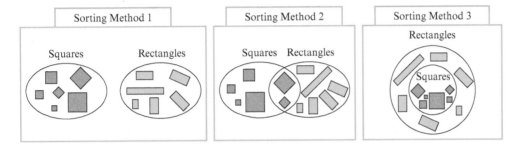

   The diagram that shows how to sort squares and rectangles *according to their definitions* is a Venn diagram for the categories of squares and rectangles.

2. Draw and explain a Venn diagram relating the categories of rhombuses, squares, and rectangles based on the definitions of those shapes.

3. Draw and explain a Venn diagram relating the categories of parallelograms and trapezoids based on the definitions of those shapes.

4. Our definition of a trapezoid is a quadrilateral with *at least one* pair of parallel sides. Some books define trapezoid as a quadrilateral with *exactly one* pair of parallel sides. How would the Venn diagram relating parallelograms and trapezoids be different if we used this other definition of trapezoid?

SECTION 10.4 **CLASS ACTIVITY 10-V** 𝕏

Using a Compass to Construct Triangles and Quadrilaterals

CCSS   CCSS SMP3, SMP5, 7.G.2

Materials You will need a compass (for drawing circles) and ruler for this activity.

Focus on the definition of a circle throughout the activity.

1. Use a compass to help you draw an isosceles triangle. Without measuring any side lengths, explain why your triangle must be isosceles.

2. Try to draw an equilateral triangle by using only a ruler and pencil, no compass. Why does this not work so well?

3. To construct an equilateral triangle, follow the steps outlined in the figure below. Notice that you really need to draw only the top portions of the circles. Use this method to make examples of several different equilateral triangles.

   Explain *why* this method must always produce an equilateral triangle.

| Step 1: Start with any line segment AB. | Step 2: Draw a circle centered at A, passing through B. | Step 3: Draw a circle centered at B, passing through A. | Step 4: Label one of the two points where the circles meet C. Connect A, B, and C with line segments. |

4. In the figure below, the line segments AB and AC have the same length. Use a ruler and compass to construct a rhombus that has AB and AC as two of its sides.

Hint: To create the rhombus, you will need to construct a fourth point, D, such that the distance from D to B is equal to the distance from D to C, and such that these two distances are also equal to the common distance from A to B and A to C. Think about the *definition of circles* to help you figure out how to construct the point D.

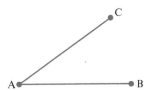

5. The line segment AB shown in the figure below is 4 inches long. Use a ruler and compass to construct a triangle that has AB as one of its sides, has a side that is 3 inches long, and has another side that is 2 inches long. Describe how you constructed your triangle, and explain why your construction must produce the desired triangle. *Hint*: Modify the construction of an equilateral triangle shown in part 3 by drawing circles of different radii.

6. Take a blank piece of paper. Use a ruler and compass to construct a triangle that has one side of length 6 inches, one side of length 5 inches, and one side of length 3 inches.

7. Is it possible to make a triangle that has one side of length 6 inches, one side of length 3 inches, and one side of length 2 inches? Explain.

SECTION 10.4 **CLASS ACTIVITY 10-W** 🍎

Making Shapes by Folding Paper

CCSS CCSS SMP3, SMP5, 7.G.2

Materials You will need paper, scissors, and a ruler for this activity.

1. To create an isosceles triangle, follow the next set of instructions.

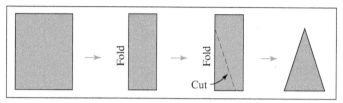

Fold a rectangular piece of paper in half. Then draw a line from the bottom edge to the folded side. Cut along this line. When you unfold, you will have an isosceles triangle.

   a. By referring to the definition of isosceles triangle, explain *why* the method described must always create an isosceles triangle.

   b. What properties does your isosceles triangle have? Find as many as you can. Explain if you can.

2. To create a rhombus, follow the next set of instructions.

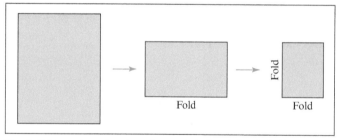

Step 1: Fold a rectangular piece of paper in half and then in half again, creating perpendicular fold lines.

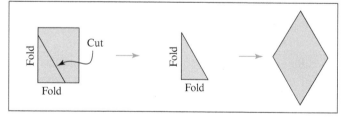

Step 2: Draw a line from anywhere on one fold to anywhere on the perpendicular fold. Cut along the line you drew. When you unfold, you will have a rhombus.

   a. By referring to the definition of rhombus, explain *why* the method described must always create a rhombus.

   b. What properties does your rhombus have? List as many as you can. Explain if you can.

3. An ordinary rectangular piece of paper is one example of a rectangle. You can create other rectangles out of an ordinary piece of paper as follows: Fold the paper so that one edge of the paper folds directly onto itself. The opposite edge will automatically fold onto itself as well. Now unfold the paper and fold the paper again, this time so that the other two edges fold onto themselves. When you unfold, you can cut along the fold lines to create four rectangles.

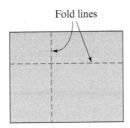

By referring to the definition of a rectangle, explain why the method just described must always create rectangles.

4. To create a parallelogram, start with a rectangular piece of paper of any size. Follow the next set of steps.

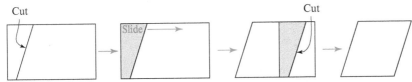

Draw a line segment connecting two opposite sides of a rectangle. Cut along the line segment. Put the piece back (shown shaded) and slide it over as shown. Mark and cut at the leading edge of the slid-over piece.

Use the converse of the Parallel Postulate to explain why the newly cut sides are parallel.

Making Shapes by Walking and Turning along Routes

CCSS   CCSS SMP1, 7.G.2, 7.G.5

1. The polygons below are regular polygons. For each polygon, give Robot Robby instructions for how to move and turn so that thier path makes the polygon. The final instruction should be for Robby to turn to face the way they started.

   To help you determine Robot Robby's angles of turning, consider their total amount of turning as they go all the way around the polygon. Which compass directions do they face? Are any repeated or omitted?

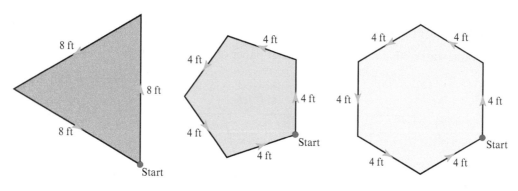

2. Give Automaton Amy instructions for how to move and turn so that their path makes a parallelogram that has a 3-meter side, a 5-meter side, and an angle of 60°. Explain how to determine the instructions.

SECTION 11.1 CLASS ACTIVITY 11-A 🍎

The Biggest Tree in the World

CCSS CCSS SMP2

Each of the trees described below could perhaps qualify as the biggest tree in the world. Compare these trees. Why can reasonable people differ about which tree is biggest?

Tree 1: General Sherman is a giant sequoia in Sequoia National Park in California. It is 275 feet tall and has a circumference (at its base) of 103 feet and a volume of 52,500 cubic feet.

Tree 2: General Grant is a giant sequoia in Sequoia National Park in California. It is 268 feet tall and has a circumference (at its base) of 108 feet and a volume of 46,600 cubic feet.

Tree 3: Mendocino tree is a redwood tree near Ukiah, California. It is 368 feet tall and has a diameter of 10.4 feet, which means that its circumference should be about 33 feet.

Tree 4: A Banyan tree in Kolkata, India, has a circumference of 1350 feet (meaning the circumference of the whole tree, not just the trunk) and covers three acres.

Tree 5: A tree in Santa Maria del Tule near Oaxaca, Mexico, is 130 feet tall and is described as requiring 40 people holding hands to encircle it.

SECTION 11.1   **CLASS ACTIVITY 11-B**

What Concepts Underlie the Process of Length Measurement?

CCSS   CCSS SMP3, SMP5, 1.MD.2

Some students were asked to measure the length of a leaf using different objects. For each example of (hypothetical) student work below, discuss ideas about length measurement that the student may not yet understand.

Student 1:

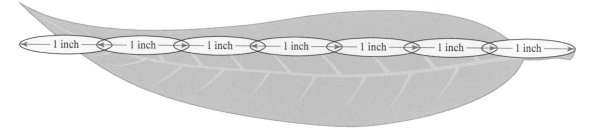

Student 2:

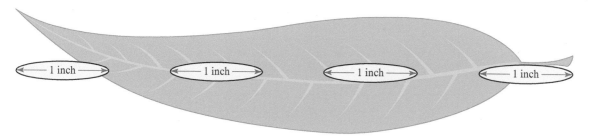

Student 3:

Student 4:

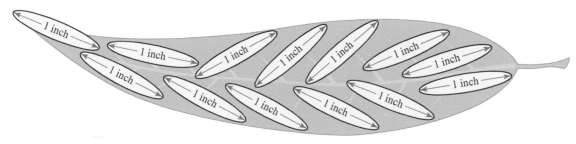

SECTION 11.1 **CLASS ACTIVITY 11-C**

Making and Using a Ruler

CCSS CCSS SMP3, SMP5, 1.MD.2, 2.MD.1

1. Show and discuss how children could make their own inch-ruler using an inch-tile and a cardboard strip like the ones shown below. What do the tick marks and numbers on the ruler mean?

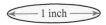

2. Children sometimes try to measure the length of an object by placing one end of the object at the 1 mark instead of the 0 mark, as shown on the centimeter ruler below. Why is the strip below not 5 cm long, even though the end of the strip is at 5? Why might a child put one end of the strip at the 1 mark? Is it possible to measure by starting at 1 or at another tick mark?

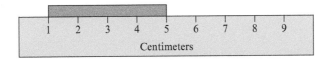

3. When asked how long the dark strip in the next figure is, some children will respond that it is 8 cm long. Others will respond that it is 7 cm long. How do you think children come up with these answers?

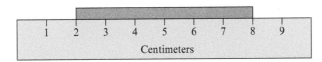

4. Some students might report that the strip measured by the inch ruler shown is 2.3 inches long. Why is this not correct? What is a correct way to report the length of the strip?

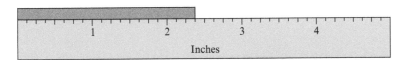

SECTION 11.1 CLASS ACTIVITY 11-D 🍎

What Does "6 Square Inches" Mean?

CCSS CCSS SMP6, 3.MD.5, 3.MD.6

Materials A set of 1-inch-by-1-inch square tiles would be helpful. Or use 1-inch graph paper from Download G-4 at bit.ly/2SWWFUX

1. Discuss the following as clearly and concretely as you can, illustrating with 1-inch-by-1-inch tiles or graph paper, if available:

   What does it mean to say that a shape has an area of 6 square inches?

   Why is it easy to think that a 6-inch-by-6-inch square has area of 6 square inches, and why is this not correct?

2. Which of the following describe the same area? Why?
   a. 4 square inches

   b. A 4-inch-by-4-inch square

   c. 4 in^2

   d. 4 in. $\times$ 4 in.

   e. A 2-inch-by-2-inch square

   f. 2 in^2

3. People sometimes say, "Area is length times width." Why is it not correct to characterize area this way?

SECTION 11.2 **CLASS ACTIVITY 11-E**

Dimension and Size

CCSS CCSS SMP4

1. Imagine a lake. Describe one-dimensional, two-dimensional, and three-dimensional parts or aspects of the lake. In each case, state how you would measure the size of that part or aspect of the lake—by length, by area, or by volume—and name an appropriate U.S. customary unit and an appropriate metric unit for measuring or describing the size of that part or aspect of the lake. What are practical reasons for wanting to know the size of these parts or aspects of the lake?

2. Imagine a house. Describe one-dimensional, two-dimensional, and three-dimensional parts or aspects of the house. In each case, state how you would measure the size of that part or aspect of the house—by length, by area, or by volume—and name an appropriate U.S. customary unit and an appropriate metric unit for measuring or describing the size of that part or aspect of the house. What are practical reasons for wanting to know the size of these parts or aspects of the house?

SECTION 11.3   CLASS ACTIVITY 11-F

Reporting and Interpreting Measurements

CCSS   CCSS SMP6

1. a. Does a food label that says "0 grams trans fat in 1 serving" mean that the food contains no trans fat? If not, what does it mean?

b. If a food label said "0.0 grams trans fat" would that mean there is no trans fat in the food?

2. One source says that the average distance from the earth to the sun is 93,000,000 miles, and another source says that the average distance from the earth to the sun is 92,960,000 miles. Can both of these descriptions be correct, or must at least one of them be wrong? Explain.

3. The label on a bottle of juice states that the bottle contains 0.5 liters of juice. To determine how many fluid ounces the juice is, Yael uses a calculator and gets the number 16.886543535620053. Discuss whether Yael should describe the amount of juice as 16.886543535620053 fluid ounces, or if not, why not, and what answer would be better.

SECTION 11.4 | **CLASS ACTIVITY 11-G** 🍎

Conversions: When Do We Multiply?
When Do We Divide?

CCSS CCSS SMP3, 4.MD.1, 5.MD.1

1. Julie is confused about why we *multiply* by 3 to convert 6 yards to feet. She thinks we should *divide* by 3 because feet are smaller than yards.

 a. Make a math drawing to show how yards and feet are related. Take care that your drawing accurately portrays length as a one-dimensional, not a two-dimensional, attribute. Use your drawing and what multiplication means to explain why we multiply by 3 to convert 6 yards to feet.

 b. Discuss the relationship between the *size* of a unit and the *number* of units it takes to describe the length of an object.

 c. Try to think of other ways to discuss conversions. What problems or questions could you pose?

2. Nate is confused about why we *divide* by 100 to convert 200 centimeters to meters. He thinks we should *multiply* by 100 because meters are bigger than centimeters. Use several approaches to discuss converting 200 centimeters to meters.

SECTION 11.4 CLASS ACTIVITY 11-H

Reasoning about Multiplication and Division to Solve Conversion Problems

CCSS   CCSS SMP1, 5.MD.1, 6.RP.3d

1. Shaquila is 57 inches tall. How tall is Shaquila in feet?

   Should you multiply or divide to solve this problem? Explain. Describe a number of different correct ways to write the answer to the conversion problem. Explain briefly why these different ways of writing the answer mean the same thing.

2. Carlton used identical paper clips to measure the length of a piece of wood. He found that the wood is 35 paper clips long. Next, Carlton measured that 2 rods are as long as 5 paper clips. How many rods long is the wood? Explain your reasoning.

3. Suppose that the students in your class want to have a party at which they will serve punch to drink. The punch that the children want to serve is sold in half-gallon containers. If 25 children attend the party and if each drinks 8 fluid ounces of punch, then how many containers of punch will you need? Describe several different ways that students could correctly solve this problem. For each method of solving the problem, explain simply and clearly why the method makes sense.

How and Why Does Dimensional Analysis Work to Convert Measurements?

CCSS CCSS SMP2

Methods A and B provide two ways of writing the steps for converting 25 meters to yards with dimensional analysis, using the fact that 1 in. = 2.54 cm.

Method A

$$\frac{25 \text{ m} \quad | \quad 100 \text{ cm} \quad | \quad 1 \text{ in.} \quad | \quad 1 \text{ ft} \quad | \quad 1 \text{ yd}}{\qquad\quad | \quad 1 \text{ m} \quad | \quad 2.54 \text{ cm} \quad | \quad 12 \text{ in.} \quad | \quad 3 \text{ ft}} \approx 27.3 \text{ yd}$$

Method B

$$25 \text{ m} \times \frac{100 \text{ cm}}{1 \text{ m}} \times \frac{1 \text{ in.}}{2.54 \text{ cm}} \times \frac{1 \text{ ft}}{12 \text{ in.}} \times \frac{1 \text{ yd}}{3 \text{ ft}} \approx 27.3 \text{ yd}$$

To carry out the calculations for method A, multiply the numbers in the top of the table and divide by the numbers in the bottom of the table. To carry out the calculations for method B, multiply the fractions.

1. **a.** Compare methods A and B.

   b. Method B works by starting with a measurement (such as 25 m) and repeatedly multiplying by certain fractions. Discuss the fractions that you multiply by. How are they chosen?

   c. Explain why method B works; in other words, explain why 25 meters really must be (approximately) equal to 27.3 yards. What is special about the fractions you multiply by that allows you to deduce this?

   d. Explain how to convert 25 meters to yards by reasoning with the meaning of multiplication and division. Then compare these calculations with the calculations you do with dimensional analysis. What do you notice?

2. Use dimensional analysis to convert 1 mile to kilometers, using the fact that 1 in. = 2.54 cm.

SECTION 11.4 CLASS ACTIVITY 11-J

Reasoning about Area and Volume Conversions

CCSS CCSS SMP2

1. If 1 yard is equal to 3 feet, does this mean that 1 square yard is 3 square feet? Make a drawing to show how many square feet are in a square yard.

2. A rug is 5 yards long and 4 yards wide. What is the area of the rug in square yards? What is the area of the rug in square feet? Show two different ways to solve this problem. Explain each case.

3. A room has a floor area of 35 square yards. What is the floor area of the room in square feet? Explain your answer.

4. A compost pile is 2 yards high, 2 yards long, and 2 yards wide. Does this mean that the compost pile has a volume of 2 cubic yards? Explain.

5. Determine the volume in cubic feet of the compost pile described in the previous question in two different ways. Explain each case.

SECTION 11.4 **CLASS ACTIVITY 11-K**

Area and Volume Conversions: Which Are Correct, and Which Are Not?

CCSS CCSS SMP3

1. Analyze the calculations that follow, which are intended to convert 25 square meters to square feet. Which use legitimate methods and are correct, and which are not? Explain.

a. $25 \text{ m}^2 = 25 \text{ m} \times \dfrac{100 \text{ cm}}{1 \text{ m}} \times \dfrac{1 \text{ in.}}{2.54 \text{ cm}} \times \dfrac{1 \text{ ft}}{12 \text{ in.}} \approx 82 \text{ ft}^2$

b. $25 \text{ m}^2 = 25 \text{ m}^2 \times \dfrac{100 \times 100 \text{ cm}^2}{1 \text{ m}^2} \times \dfrac{1 \text{ in.}^2}{2.54 \times 2.54 \text{ cm}^2} \times \dfrac{1 \text{ ft}^2}{12 \times 12 \text{ in.}^2} \approx 269 \text{ ft}^2$

c. $25 \text{ m} = 25 \text{ m} \times \dfrac{100 \text{ cm}}{1 \text{ m}} \times \dfrac{1 \text{ in.}}{2.54 \text{ cm}} \times \dfrac{1 \text{ ft}}{12 \text{ in.}} \approx 82 \text{ ft}$

Therefore,

$$25 \text{ m}^2 \approx 82^2 \text{ ft}^2 = 6727 \text{ ft}^2$$

d. 25 square meters is the area of a square that is 5 meters wide and 5 meters long, so

$$5 \text{ m} = 5 \text{ m} \times \dfrac{100 \text{ cm}}{1 \text{ m}} \times \dfrac{1 \text{ in.}}{2.54 \text{ cm}} \times \dfrac{1 \text{ ft}}{12 \text{ in.}} \approx 16.404 \text{ ft}$$

Therefore,

$$25 \text{ m}^2 \approx 16.404 \times 16.404 \text{ ft}^2 \approx 269 \text{ ft}^2$$

2. Use the fact that 1 in. = 2.54 cm to convert 27 cubic feet to cubic meters in at least two different ways.

Model with Conversions

CCSS   CCSS SMP4

1. How much water would you expect the people in a city with a population of 100,000 to use in 1 day? What size tank would hold this amount of water? Compare this size container with something familiar.

To answer these questions, first estimate how much water each person might use per day. Then do some calculations. Consider working with metric measurements.

Note that many showers have a water flow of 8 liters per minute and many toilets use about 5 liters of water per flush.

2. How many square miles of land would you need to plant 1 million trees? 1 billion trees? 1 trillion trees? Compare these land areas to familiar areas. You may wish to assume you can plant trees in rows that are 10 feet apart and that the trees are 10 feet apart in each row.

SECTION 12.1 CLASS ACTIVITY 12-A

Units of Length and Area in the Area Formula for Rectangles

CCSS CCSS SMP2, 3.MD.5, 3.MD.6, 3.MD.7, 5.NF.4b

Materials If available, square centimeter tiles would be helpful.

1. What does it mean to say that a shape has an area of 15 square centimeters?

2. The large rectangle shown here is 3 cm by 5 cm. What is a direct way to determine the area of the rectangle in square centimeters that relies on the meaning of area?

A 1-cm-by-
1-cm square
of area 1 cm^2

3. In Chapter 4, we defined multiplication in terms of equal groups. According to our definition, $3 \cdot 5$ means the number of units in 3 groups of 5 units. Using our definition of multiplication, explain why the area of the large rectangle in part 2 is $3 \cdot 5$ cm^2.

4. The *length times width* area formula for rectangles involves lengths, but doesn't explicitly involve equal groups. Discuss the following:

    How are the *lengths* 3 cm and 5 cm *linked to* yet *different from* the numbers of groups and numbers of things in each group in your explanation for part 3?

5. **a.** Given that the line segment shown is 1 unit long, use the grid to lightly shade a rectangle that is $\frac{7}{10}$ units by $\frac{9}{10}$ units.

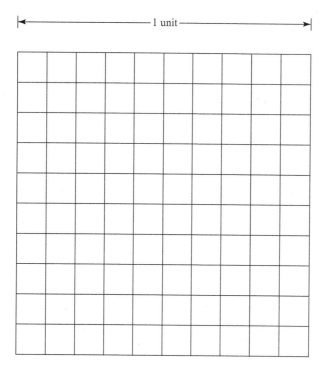

|← ———————— 1 unit ———————— →|

    b. Apply the *length times width* formula to find the area of the shaded rectangle in part (a) and verify that the formula gives you the correct area for your rectangle. Attend carefully to units of area.

    c. Show and describe the $\frac{7}{10}$- and $\frac{9}{10}$-unit lengths on your rectangle, keeping in mind that length is a one-dimentional attribute. Explain why it would be confusing to say that the small squares represent these lengths.

SECTION 12.2   CLASS ACTIVITY 12-B

How Can We Create Different Shapes of the Same Area?

CCSS CCSS SMP1, SMP3, 3.MD.5b, 3.MD.6

Materials You will need Download 12-1 at bit.ly/2SWWFUX and scissors and tape.

1. Cut out the two 4-unit-by-6-unit rectangles from the Download. Leave one rectangle as it is. Use the other rectangle to make another shape that has the exact same area. Feel free to be creative! Make a rough sketch of the new shape you created.

   A 4-unit-by-6-unit rectangle: Sketch of your new shape of the same area:

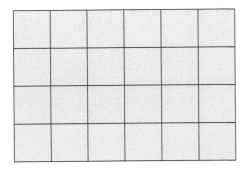

2. Why does the new shape you created have the same area as the 4-unit-by-6-unit rectangle? Try to find more than one way to explain!

3. In contrast with part 1, what could you do to the 4-unit-by-6-unit rectangle to create a shape of a *different* area? Give some examples.

SECTION 12.2 **CLASS ACTIVITY 12-C** 🍎

Reasoning with the Moving and Additivity Principles

CCSS CCSS SMP3, 3.MD.7d

Materials Optionally, use Downloads 12-2 and 12-3 at bit.ly/2SWWFUX and scissors and tape.

1. Use the moving and additivity principles (one or both) to determine the area of the L-shaped region in several different ways. In each case, explain your reasoning and write and evaluate an algebraic expression that fits with the strategy. Try to find strategies of the following types:

   • A simple subdividing strategy

   • A takeaway strategy

   • A move and reattach strategy

   • A combine two copies and take half strategy

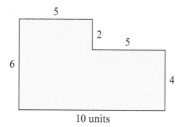

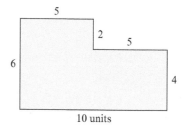

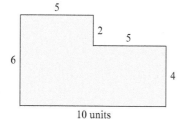

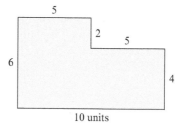

2. Use the moving and additivity principles (one or both) to determine the area of the shaded square, which is inside the 2-unit-by-2-unit square, in several different ways. In each case, explain your reasoning.

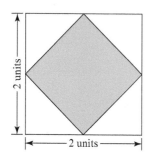

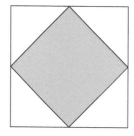

SECTION 12.3 CLASS ACTIVITY 12-D

Determining Areas of Triangles in Progressively Sophisticated Ways

CCSS CCSS SMP2, 6.G.1

Materials Optionally, use Downloads 12-4 or 12-5 and 12-6 at bit.ly/2SWWFUX and scissors and tape.

1. Use the moving and additivity principles to determine the area of the next triangles in three different ways: (a) by moving small pieces and relying directly on the definition of area, (b) by moving bigger chunks to create a rectangle, and (c) by viewing the triangle as part of a larger rectangle. Use the 3 copies of each triangle to show the 3 different ways.

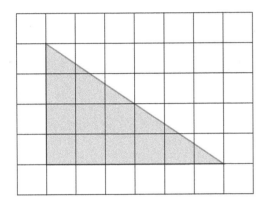 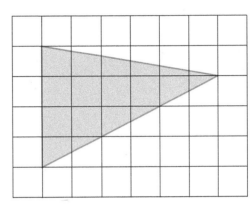

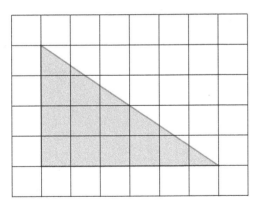 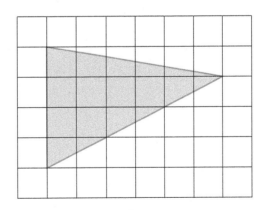

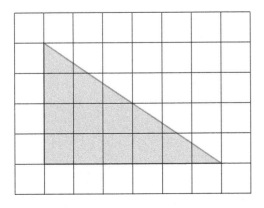 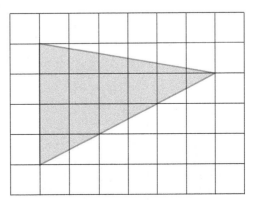

2. Think about some of the methods you used to determine the areas of the triangles in part 1, such as moving and reattaching chunks, or making two copies and then taking half. Find arithmetic problems that can be made easy to solve by using numerical strategies that are similar to these geometric strategies. Describe how to solve these arithmetic problems and say briefly how the solution methods are roughly similar to the geometric methods.

Choosing the Base and Height of Triangles

CCSS   CCSS SMP5, SMP7

Use the three copies of the triangle below to show the three different ways to choose the base and height of the triangle. Once you have chosen a base, the right angle formed by the corner of a piece of paper may help you determine where to draw the height, which must be perpendicular to the base (or an extension of the base).

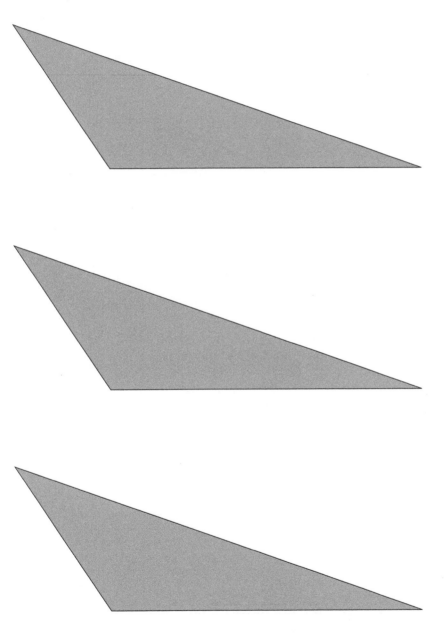

SECTION 12.3 **CLASS ACTIVITY 12-F** 🍎

Explaining Why the Area Formula for Triangles Is Valid

CCSS CCSS SMP3, 6.G.1

Materials Optionally, use Downloads 12-7 and 12-8 at bit.ly/2SWWFUX and scissors and tape.

1. Use the moving and additivity principles to explain in three ways why the next triangle has area $\frac{1}{2}(b \cdot h)$ square units for the given choices of b and h. One explanation should fit naturally with the expression $\frac{1}{2}(b \cdot h)$, another explanation should fit naturally with the expression $(\frac{1}{2}b) \cdot h$, and a third explanation should fit naturally with the expression $b \cdot (\frac{1}{2}h)$. Why is it valid to describe the area with any one of these three expressions?

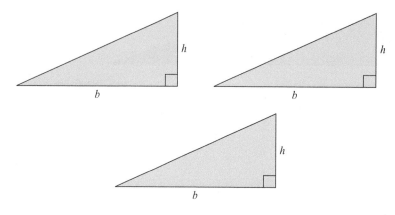

2. Use the moving and additivity principles to explain in two ways why the next triangle has area $\frac{1}{2}(b \cdot h)$ square units for the given choices of b and h. One explanation should fit naturally with the expression $b \cdot (\frac{1}{2}h)$ and the other should fit naturally with the expression $\frac{1}{2}(b \cdot h)$.

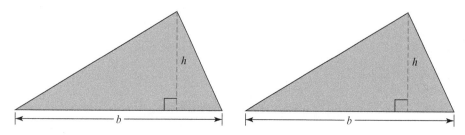

3. What is wrong with the following reasoning that claims to show that the area of the triangle ABC below is $\frac{1}{2}(b \cdot h)$ square units?

Draw a rectangle around the triangle ABC, as shown on the right in the figure below. The area of this rectangle is $b \cdot h$ square units. The line AC cuts the rectangle in half, so the area of the triangle ABC is half of $b \cdot h$ square units—in other words, $\frac{1}{2}(b \cdot h)$ square units.

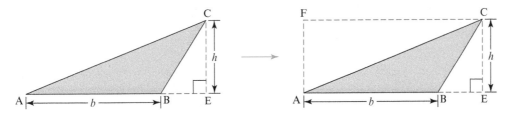

4. What is a valid way to explain why the shaded triangle in part 3 has area $\frac{1}{2}(b \cdot h)$ square units for the given choice of b and h?

Suggestion: Let a be the length of the line segment from B to E. Because we have already explained why the triangle formula is valid for *right triangles*, you may apply it to right triangles in the figure.

SECTION 12.3 CLASS ACTIVITY 12-G

Area Problem Solving

CCSS CCSS SMP1, SMP3

1. Determine the area of the shaded triangle that is inside the rectangle. Explain your reasoning.

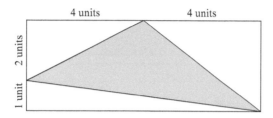

2. Determine the area of the shaded triangle that is in the rectangle in *two different ways.* Explain your reasoning in each case.

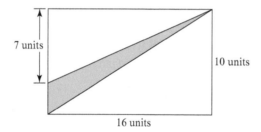

3. Determine the area of the shaded triangle in the figure below in *two different ways.* Explain your reasoning in each case.

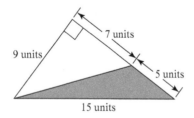

4. Determine the area of the shaded triangle below. Explain your reasoning.

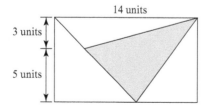

SECTION 12.4 | **CLASS ACTIVITY 12-H**

Do Side Lengths Determine the Area of a Parallelogram?

CCSS CCSS SMP8, 6.G.1

1. The three parallelograms below (the first of which is also a rectangle) all have two sides that are 3 units long and two sides that are 7 units long.

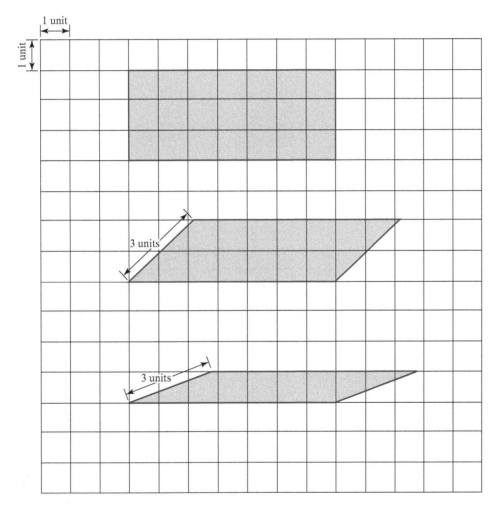

   a. Use the moving and additivity principles to determine the areas of the three parallelograms.

   b. Can there be a formula for areas of parallelograms that is only in terms of the lengths of the sides? Explain why or why not.

2. Find a formula for the area of a parallelogram *in terms of lengths of parts of the parallelogram.* Use the following parallelogram to help you describe your formula:

SECTION 12.4 **CLASS ACTIVITY 12-I**

Explaining Why the Area Formula for Parallelograms Is Valid

CCSS CCSS SMP3, 6.G.1

Materials Optionally, use Download 12-9 at bit.ly/2SWWFUX and scissors and tape.

1. Show how to subdivide and recombine the parallelogram below to form a b by h rectangle, thereby explaining why the area of the parallelogram is $b \cdot h$.

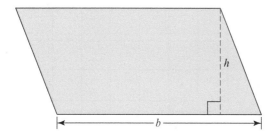

2. Explain why the area of the shaded parallelogram below is $b \cdot h$. To do so, consider using these approaches: (a) enclose the parallelogram or (b) subdivide the parallelogram. You may use area formulas that we have already explained.

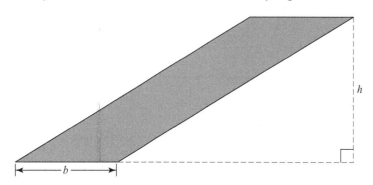

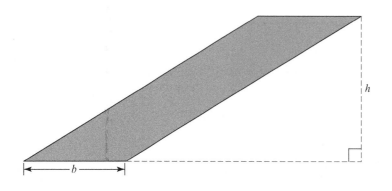

Finding and Explaining a Trapezoid Area Formula

CCSS   CCSS SMP3, 6.G.1

Materials Optionally, use Download 12-10 at bit.ly/2SWWFUX and scissors and tape.

Use several different methods to find and explain a formula for the area of a trapezoid that has parallel sides of length a and b and height h. Consider these ideas as well as others: (a) Combine two copies of the trapezoid to make a parallelogram. (b) Subdivide the trapezoid into *two* triangles, one with base b and one with base a.

If some of your formulas look different; explain why they are equivalent.

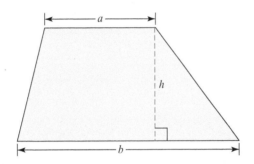

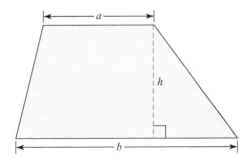

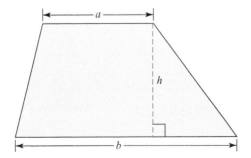

SECTION 12.5   CLASS ACTIVITY 12-K

Is This Shearing?

CCSS CCSS SMP2

The figure below shows a rectangle made of toothpicks being sheared into a parallelogram. During the process of shearing, which of the following change and which remain the same?

- Lengths of the sides
- Height of the stack of toothpicks
- Area

The figure below shows a rectangle made with pieces of drinking straws tied with a string being "squashed" into a parallelogram. During the process of squashing, which of the following change and which remain the same?

- Lengths of the sides
- "Vertical height" of the straw figure
- Area

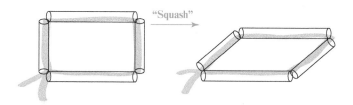

Is the process of squashing the same as shearing? Why or why not?

SECTION 12.5 **CLASS ACTIVITY 12-L**

Solving Problems by Shearing

Solve the following problems by finding and shearing appropriate triangles. To shear your triangles, identify a base and shear *parallel to the base*.

CCSS CCSS SMP1

1. Triangle ABC is divided into a triangle, which has area T, and another region, which has area R. Show and explain how to use shearing to divide triangle ABC into *two* triangles, one of area T and one of area R.

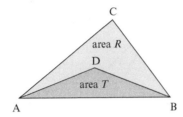

2. Show and explain how to use shearing to create a triangle that has the same area as the shaded region EFGH below.

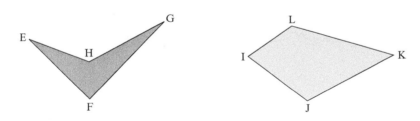

3. Show and explain how to use shearing to create a triangle that has the same area as the shaded region IJKL above.

SECTION 12.6 CLASS ACTIVITY 12-M

How Are the Circumference and Diameter of a Circle Related, Approximately?

CCSS CCSS SMP2

The circle below has diameter 1 unit. Notice that 6 equilateral triangles fit inside the circle and a square surrounds the circle. What does that suggest about the circumference of the circle? In particular, what are two numbers that the circumference lies between? Which of those two numbers does the circumference seem to be closer to?

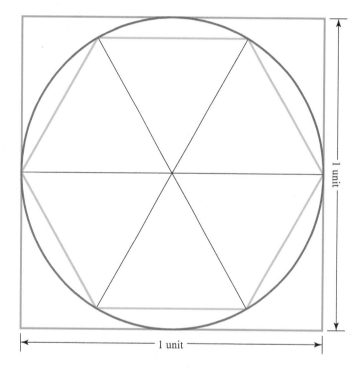

SECTION 12.6 **CLASS ACTIVITY 12-N** 🍎

How Many Diameters Does It Take to Make the Circumference of a Circle?

CCSS CCSS SMP2, SMP5, 7.G.4

Materials You will need Download 12-11 at bit.ly/2SWWFUX and scissors for this activity.

1. Cut out the circumference strips from the Download. By wrapping the edge of each circumference strip around its circle, verify that the length of each circumference strip really is the circumference of its circle.

2. Use Circle 1's diameter to measure the circumference strip for Circle 1. How many diameters does it take to make the circumference? Then use Circle 2's diameter to measure Circle 2's circumference strip and use Circle 3's diameter to measure Circle 3's circumference strip. What do you notice about all these measurements?

3. Based on this experiment, if the diameter of a circle is D centimeters and its circumference is C centimeters, how do you expect D and C to be related? Explain!

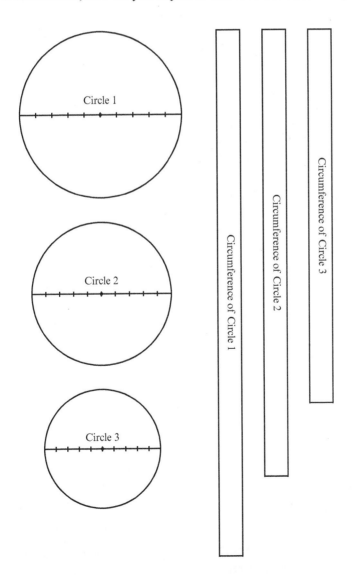

SECTION 12.6 CLASS ACTIVITY 12-O 🍎

Where Does the Area Formula for Circles Come From?

CCSS CCSS SMP3, 7.G.4

Materials You will need scissors for parts 1 and 2 (which are optional). You can also use Download 12-12 at bit.ly/2SWWFUX

1. On separate paper, draw a large circle, shade or color half of it, cut it into 8 "pie pieces," and arrange the pieces as shown below.

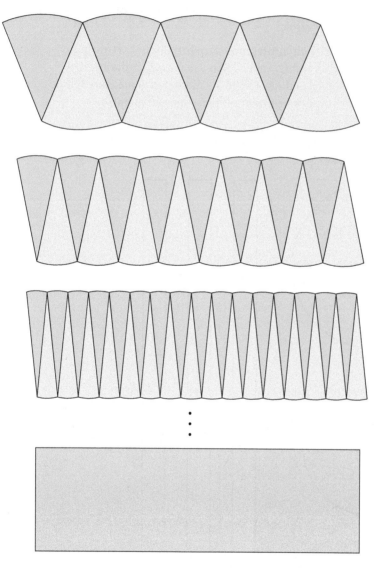

2. Now cut your circle into 16 "pie pieces." Arrange the 16 pieces as shown above.

3. Imagine cutting a circle into more and more smaller and smaller pie pieces and rearranging them as above.

   • What shape would your rearranged circle become more and more like?

   • What would the lengths of the sides of this shape be?

   • What would the area of this shape be?

4. Using your answers to part 3, explain why it makes sense that a circle of radius r units has area πr^2 square units, given that the circumference of a circle of radius r is $2\pi r$.

SECTION 12.6 **CLASS ACTIVITY 12-P**

Area Problems

CCSS CCSS SMP1, 7.G.4

Materials Optionally, for part 2, simulate the situation with string and a box.

1. A reflecting pool will be made in the shape of the shaded region shown in the figure below. The arcs shown are from quarter-circles. What is the area of the surface of the pool? Explain your reasoning.

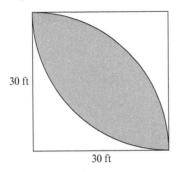

30 ft

30 ft

2. Melchior the goat will be tied by a 15 meter rope to the outside of a 10-meter-by-16-meter rectangular shed that is surrounded by grass. Try three different places to attach the rope. For each, determine the area of grass that Melchior could eat. Which of your locations gives Melchior the most to eat? (Note that Melchior cannot walk into or through the shed.)

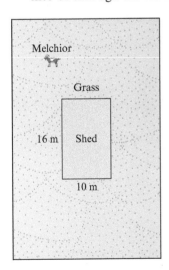

Melchior

Grass

16 m   Shed

10 m

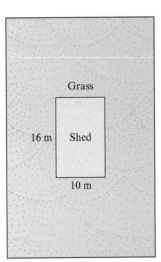

Grass

16 m   Shed

10 m

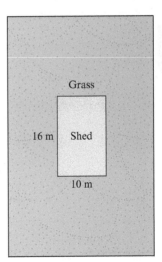

Grass

16 m   Shed

10 m

3. Given a regular octagon that has perimeter *P* units and whose distance from the center to a side is *r* units, find a formula in terms of *P* and *r* for the area of the octagon. Explain your reasoning. Is this similar to the case of a circle?

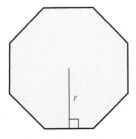

r

SECTION 12.7 **CLASS ACTIVITY 12-Q**

Determining the Area of an Irregular Shape

CCSS CCSS SMP5

Materials String and some of the other items in the bullets below would be helpful. Optionally, use Download 12-13 at bit.ly/2SWWFUX

1. Think about several different ways that you might determine approximately the area of the surface of Lake Lalovely shown on the map below. Suppose that you have the following items on hand:

   • Many 1-inch-by-1-inch plastic squares

   • Graph paper (adjacent lines separated by $\frac{1}{4}$ inch)

   • A scale for weighing (such as one used to determine postage)

   • String

   • Modeling dough

   Which of these items could help you to determine approximately the area of the surface of the lake? How?

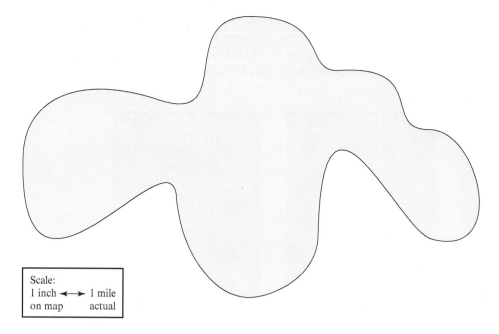

Scale:
1 inch ⟷ 1 mile
on map   actual

2. Suppose that you have a map with a scale of 1 inch ⟷ 100 miles. You trace a state on the map onto $\frac{1}{4}$-inch graph paper. (The grid lines are spaced $\frac{1}{4}$ inch apart.) You find that the state takes up about 108 squares of graph paper. Approximately what is the area of the state? Explain.

3. Some students were working on the problem in part 2 and made the initial calculations shown below. For each of these initial calculations, either explain the ideas that could be behind the calculations and use them to finish determining the area of the state, or explain how the calculations might lead to an incorrect area for the state.

   a. $100 \div 4 = 25$ $25 \cdot 25 = 625$

   b. $108 \div 4 = 27$ $27 \cdot 100 = 2700$

   c. $4 \cdot 4 = 16$ $108 \div 16 = 6.75$

   d. $\frac{1}{4} \cdot \frac{1}{4} = \frac{1}{16}$ $108 \cdot \frac{1}{16} =$

   Now write a numerical expression for the area of the state and relate it to the correct calculation methods.

4. Suppose that you have a map with a scale of 1 inch $\leftrightarrow$ 15 miles. You cover a county on the map with a $\frac{1}{8}$-inch-thick layer of modeling dough. Then you re-form this piece of modeling dough into a $\frac{1}{8}$-inch-thick rectangle. The rectangle is $2\frac{1}{2}$ inches by $3\frac{3}{4}$ inches. Approximately what is the area of the county? Explain.

5. Suppose that you have a map with a scale of 1 inch $\leftrightarrow$ 50 miles. You trace a state onto cardstock. Using a scale, you determine that a full $8\frac{1}{2}$-inch-by-11-inch sheet of cardstock weighs 10 grams. Then you cut out and weigh the cardstock tracing of the state. It weighs 6 grams. What is the approximate area of the state? Explain.

SECTION 12.8 | CLASS ACTIVITY 12-R 🍎

Critique Reasoning about Perimeter

CCSS CCSS SMP3, 3.MD.8

Materials You will need a centimeter ruler for part 2.

1. Johnny's method for calculating the perimeter of the shaded shape in Figure (a) is to shade the squares along the border of the shape, as shown in Figure (b), and to count these border squares. Therefore, Johnny says the perimeter of the shape is 24 units. Is Johnny's method valid? If not, why not?

(a) 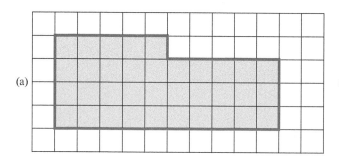   (b)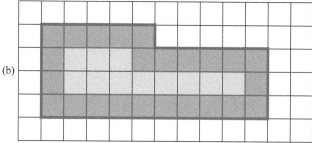

2. When Susie was asked to draw a shape with perimeter 15 cm, she drew a shape like the shaded one shown in the figure below on centimeter grid paper.
   a. Carefully measure the diagonal line segment in the shaded shape with a centimeter ruler. Then explain why the shape does not have perimeter 15 cm.

   b. Draw a shape that has perimeter 15 cm on the same graph paper. (The vertices of your shape do not have to be located where grid lines meet.) Explain how you figured out how to make your shape.

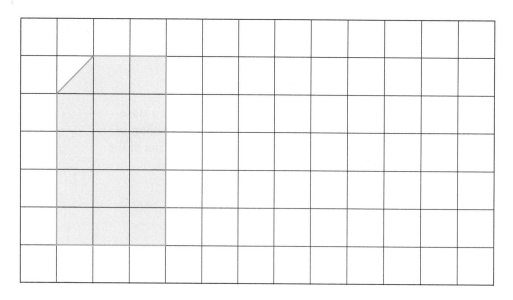

SECTION 12.8 **CLASS ACTIVITY 12-S**

Find, Explain, and Use Perimeter Formulas for Rectangles

CCSS CCSS SMP7, 4.MD.3

1. Describe several different methods for determining the perimeter of a rectangle. For each method, write the corresponding formula for the perimeter P of an A-unit-by-B-unit rectangle.

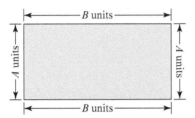

2. Find 3 different rectangles that have perimeter 18 centimeters.

3. Now find at least 3 more different rectangles that have perimeter 18 centimeters, including some whose side lengths are not whole numbers.

4. Discuss: Is there a systematic way to describe and find all rectangles of perimeter 18 centimeters? Is one of your formulas from part 1 especially helpful for this purpose?

SECTION 12.8 CLASS ACTIVITY 12-T 🍎

How Are Perimeter and Area Related for Rectangles?

CCSS CCSS SMP8, 3.MD.8, 4.MD.3

Materials String would be useful for this activity. You might also like to use Download 12-14 at bit.ly/2SWWFUX and scissors.

1. If a rectangle has perimeter 20 units, then what could its area be? Draw at least 5 different rectangles of perimeter 20 units on graph paper (and cut them out if you like). Include some rectangles that have sides whose lengths *aren't whole numbers of units.* Describe a strategy for finding rectangles with perimeter 20 units.

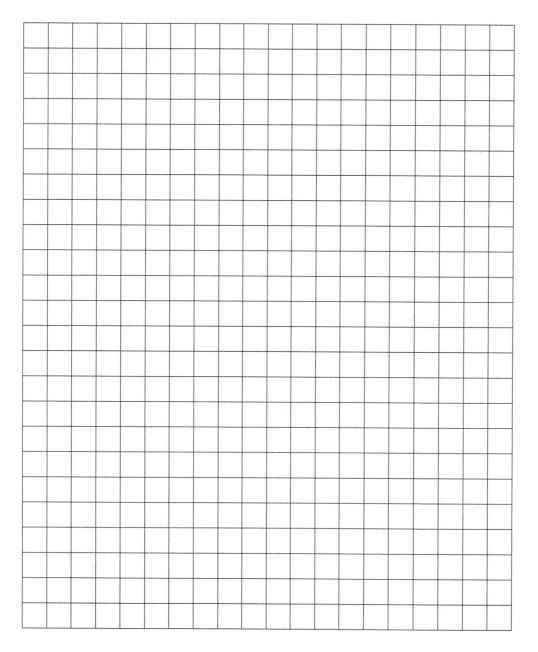

2. Find the area of each of your rectangles in part 1. In the table below, list the areas of your rectangles from part 1 in decreasing order. Below each area, draw a small sketch showing the approximate shape of the corresponding rectangle.

How are the larger-area rectangles qualitatively different in shape from the smaller-area ones?

Greater area   Smaller area

Area								
Shape								

3. Show how two people can use a loop of string and 4 fingers to represent all rectangles of a certain fixed perimeter.

Now consider all the rectangles of perimeter 20 units, *including those whose side lengths aren't whole numbers of units.* What are all the theoretical possibilities for the areas of those rectangles? What is the largest possible area, and what is the smallest possible area (if there is one)?

SECTION 12.8 **CLASS ACTIVITY 12-U**

How Are Perimeter and Area Related for All Shapes?

CCSS CCSS SMP3, SMP5

Materials String would be helpful for this activity.

1. Nick wants to find the area of the irregular shape below. He cuts a piece of string so that it goes all the way around the outside of the shape and then forms his piece of string into a square on top of graph paper. Using the graph paper, Nick gets a good estimate for the area of his string square and then uses the square's area as his estimate for the area of the original irregular shape.

   Discuss whether Nick's method is a legitimate way to estimate areas of irregular shapes.

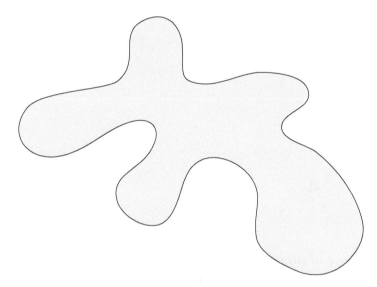

2. Suppose that a forest on flat terrain has perimeter 200 kilometers, but there is no information on the shape of the forest. What can you say about the forest's area? Use a loop of string to help you think about this question. Then consider these questions:

   • What shape do you think would give the largest possible area for the forest? What is this area?

   • What range of areas do you think are possible for the forest?

Side Lengths of Squares inside Squares

CCSS CCSS SMP7, 8.EE.2

Throughout this activity, assume that you don't yet know the Pythagorean theorem.

1. Use the moving and additivity principles (one or both) to determine the area of the shaded "tilted" square inside square (a) in part 2 below. Then use the area of the tilted square to determine its side lengths. Repeat with square (b).

2. Draw your own "tilted squares" inside squares (c) and (d) and repeat the instructions for part 1.

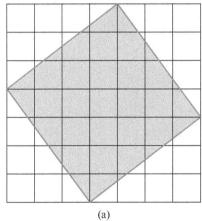

(a)

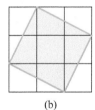

(b)

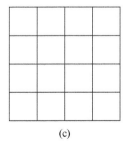

(c)

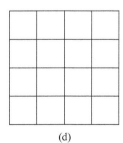

(d)

SECTION 12.9 | CLASS ACTIVITY 12-W 🍎

A Proof of the Pythagorean Theorem

CCSS   CCSS SMP3, 8.G.6

Materials Optionally, use scissors and Download 12-15 at bit.ly/2SWWFUX for part 1.

Starting with any arbitrary right triangle, like the one below, we must explain why $a^2 + b^2 = c^2$ is true.

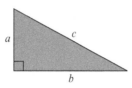

1. Follow the instructions below to show two different ways of filling a square that has sides of length $a + b$ with triangles and squares without gaps or overlaps. You may wish to cut out the squares and triangles on the Download.

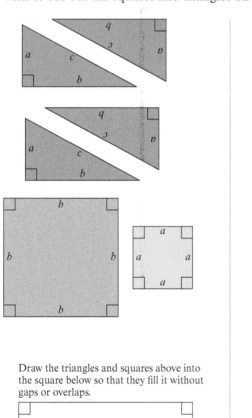

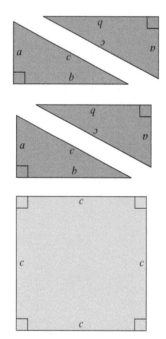

Draw the triangles and squares above into the square below so that they fill it without gaps or overlaps.

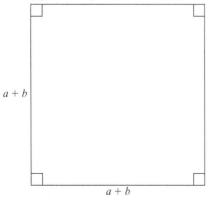

Draw the triangles and square above into the square below so that they fill it without gaps or overlaps.

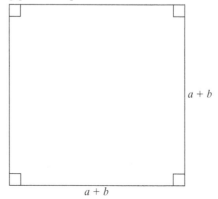

2. Now use part 1 and the moving and additivity principles about areas to explain why $a^2 + b^2 = c^2$. There are several different ways to do this.

   Hint: Notice that *both* of the two ways of filling a square of side length $a + b$ use 4 copies of the original right triangle.

   Summarize your proof of Pythagoras's theorem.

3. Here is a subtle point in the proof of the Pythagorean theorem: In both of the squares of side length $a + b$ on the previous page, there are places along the edges where 3 figures (2 triangles and 1 square) meet at a point. How do we know that the edge formed there really is a straight line and doesn't actually have a small bend in it, such as pictured in the figure below? We need to know that the edge there really is straight in order to know that the assembled shapes really do create large *squares* and not *octagons*. Explain why these edges really are straight. (*Hint:* Consider the angles at the points where a square and two triangles meet.)

4. Here is another subtle point in the proof of the Pythagorean theorem. Why does the proof explain the theorem for *all* right triangles? After all, we used one specific right triangle to explain the proof.

SECTION 13.1 **CLASS ACTIVITY 13-A**

Making Prisms and Pyramids

CCSS CCSS SMP7, K.G.5

Materials You will need toothpicks or straws and gumdrops or modeling clay for this activity. If it is available, you could also use liquid soap.

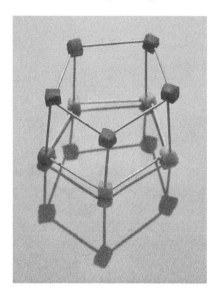

Make the shapes listed below. In each case, visualize the shape first and predict how many toothpicks and clay balls you will need to make it.

1. Rectangular prism
2. Triangular prism
3. Pyramid with a triangle base
4. Pyramid with a square base

Some teachers like to dip these shapes into a liquid soap solution to show the faces of the shapes.

| SECTION 13.1 | CLASS ACTIVITY 13-B |

Analyzing Prisms and Pyramids

CCSS CCSS SMP7, SMP8

1. Answer the questions below *without* using a model. Use your visualization skills and look back at other models. (When you are done, verify with a model if one is available.)

   • How many faces does a pentagonal prism have? Why? What kinds of shapes are they? How many of each kind of shape are there?

   • How many edges and how many vertices does a pentagonal prism have? Explain.

2. Answer the questions below *without* using a model. Use your visualization skills and look back at other models. (When you are done, verify with a model if one is available.)

   • How many faces does a pyramid with a hexagonal base have? Why? What kinds of shapes are they? How many of each kind of shape are there?

   • How many edges and how many vertices does a pyramid with a hexagonal base have? Explain.

3. Discuss the difference between a rectangle and a rectangular prism.

4. Discuss the difference between a triangle and a triangular pyramid.

SECTION 13.1   CLASS ACTIVITY 13-C

What's inside the Magic 8 Ball?

CCSS CCSS SMP7

Materials You will need a Magic 8 Ball for parts 1 and 3 of this activity. (One or more can be shared by a class.) Scissors, tape, and Downloads 13-1, 13-2, 13-3, and 13-4 at bit.ly/2SWWFUX or snap-together plastic polygons would also be helpful.

1. There is a polyhedron inside the Magic 8 Ball. What can you tell about this polyhedron without breaking open the Magic 8 Ball?

2. Let's say someone wants to make a polyhedron to put inside a handmade Magic 8 Ball. This polyhedron would probably be very uniform and regular all the way around, so that all the answers would be equally likely to appear. Such a shape might have the following properties:

   • The faces are identical copies of one regular polygon; so all faces are equilateral triangles, or all faces are squares, or all faces are regular pentagons, and so forth.

   • The shape has no indentations or protrusions.

   • All the vertices are identical, so the same number of faces meet at each vertex.

   Make a guess: How many such shapes do you think there can be?

   Try to make some shapes that have the properties described above. Tape paper polygons together (use the Downloads) or use snap-together plastic polygons. How many shapes can you find?

3. The actual Magic 8 Ball should contain one of the shapes that has the properties described in part 2. Which one must it be?

SECTION 13.1 **CLASS ACTIVITY 13-D**

Making Platonic Solids

CCSS CCSS SMP1

Materials You will need toothpicks and modeling clay (or gumdrops) for this activity. Or use scissors, tape, and Downloads 13-1, 13-2, and 13-3 at bit.ly/2SWWFUX or snap-together plastic polygons.

Make all five Platonic solids by sticking toothpicks into small balls of clay. (Your dodecahedron may sag a little, but the others will be more stable.) Or make the Platonic solids by putting plastic polygons together or by using the Downloads. Refer to the following descriptions of the Platonic solids:

Tetrahedron has 4 equilateral triangle faces, with 3 triangles coming together at each vertex.

Cube has 6 square faces, with 3 squares coming together at each vertex.

Octahedron has 8 equilateral triangle faces, with 4 triangles coming together at each vertex.

Dodecahedron has 12 regular pentagon faces, with 3 pentagons coming together at each vertex.

Icosahedron has 20 equilateral triangle faces, with 5 triangles coming together at each vertex.

Use your models to fill in the following table:

Shape	Number and Type of Faces	Number of Edges	Number of Vertices
Tetrahedron	4 equilateral triangles		
Cube	6 squares		
Octahedron	8 equilateral triangles		
Dodecahedron	12 regular pentagons		
Icosahedron	20 equilateral triangles		

Do you notice any relationships among the numbers?

SECTION 13.2 **CLASS ACTIVITY 13-E** 🍎

What Shapes Do These Patterns Make?

CCSS CCSS SMP7, SMP8, 6.G.4

Materials Scissors, tape, and Downloads 13-5, 13-6, and 13-7 at bit.ly/2SWWFUX would be helpful.

Small patterns for solid shapes are shown below. *Visualize* the shapes that these patterns will make. What shapes are they?

Compare the patterns and note similarities. Which shapes will be related or alike, and how will these shapes be related or alike?

Use the Downloads to make the shapes and check your answers.

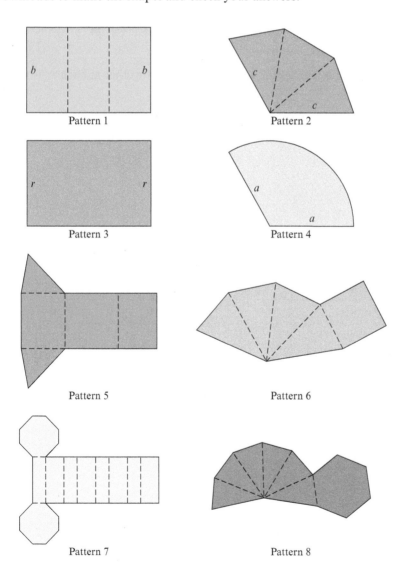

Pattern 1

Pattern 2

Pattern 3

Pattern 4

Pattern 5

Pattern 6

Pattern 7

Pattern 8

SECTION 13.2 CLASS ACTIVITY 13-F

Patterns and Surface Area for Prisms and Pyramids

CCSS CCSS SMP7, 6.G.4, 7.G.6, 8.G.7

Materials You will need scissors and centimeter graph paper (use Download G-5 at bit.ly/2SWWFUX) for parts 1 and 3 and an object in the shape of a pyramid for part 2.

1. Use graph paper to make a pattern for a prism whose bases are a rectangle that is not a square. Include the bases in your pattern.

   Let's say that the width, length, and height of your prism are W cm, L cm, and H cm respectively (it's up to you which edge length you call which). Show which lengths on your pattern are W, L, and H. Then find a formula in terms of W, L, and H for the total surface area of a prism and explain your reasoning.

   Cut out your pattern and see if it works.

2. If you have a pyramid-shaped object, use it to make a pattern for the pyramid as follows. Place the object on a piece of paper and trace around the outside of the face that is on the paper. Then repeatedly turn the object to adjacent faces and trace those faces to make the other parts of the pattern. See if you can use this method to make several different-looking patterns for the pyramid!

3. On centimeter graph paper, make a pattern for a pyramid with a 6-cm-by-6-cm square base such that the apex will be 4 cm above the center of the base. Cut out your pattern and check that it works. Then find the surface area of the pyramid.

SECTION 13.2   CLASS ACTIVITY 13-G

Patterns and Surface Area for Cylinders

CCSS CCSS SMP1, 7.G.6

Materials If available, blank paper or inch graph paper (use Downloads G-4 or G-6 at bit.ly/2SWWFUX), a ruler, scissors, and tape would be helpful.

1. Take a standard 8.5-inch-by-11-inch piece of paper, roll it up, and tape it, without overlapping the paper, to make a cylinder without bases, as shown in the figure below. What is the area of the surface of the cylinder not including the bases? Why?

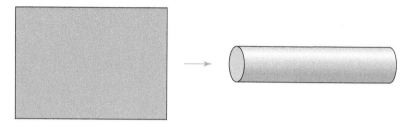

What is the surface area of the cylinder (including the bases)? Explain.

2. A company wants to manufacture tin cans that are 3 inches in diameter and 4 inches tall. Describe the shape and dimensions of the paper label the company will need to wrap around the side of each can. Explain your reasoning. Make a label of those dimensions and check that it works.

SECTION 13.2   CLASS ACTIVITY 13-H

Patterns and Surface Area for Cones

CCSS   CCSS SMP1, SMP7, 7.G.4

Materials You will need blank paper, scissors, and a compass for this activity. If available, a small cone-shaped object would be helpful for part 1.

1. How can you make a pattern for the lateral portion (i.e., the side portion, which does not include the base) of a cone? The pattern should not require the paper to overlap. Describe which edges would be joined to make the cone from the pattern.

    If you have a cone-shaped object, try to use it to make a pattern for the lateral portion of the cone by placing the object on paper and then rolling and tracing it suitably.

2. On a piece of paper, draw half of a circle that has radius 5 inches. (*Suggestion:* Put the point of your compass in the middle of the long edge of the piece of paper.) Cut out the half-circle, and attach the two radii on the straight edge (the diameter) to form a cone without a base. Calculate the radius of the circle that will form a base for your cone. Explain your reasoning. Then draw this circle, and verify that it does form the base of the cone.

3. Make a pattern for a cone such that the base is a circle of radius 2 inches and the lateral portion of the cone is made from a half-circle. Determine the total surface area of your cone. Explain your reasoning.

4. Make a pattern for a cone such that the base is a circle of radius 2 inches and the lateral portion of the cone is made from part of a circle of radius 6 inches. What fraction of the 6-inch circle will you need to use? Determine the total surface area of your cone. Explain your reasoning.

Cross-Sections of a Pyramid

CCSS CCSS SMP7, 7.G.3

Materials Modeling clay and dental floss would be helpful for part 2.

The following math drawing shows a pyramid with a square base like an Egyptian pyramid:

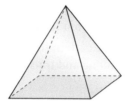

1. Visualize a plane slicing through the pyramid. The places where the plane meets the pyramid form a shape in the plane. Which shapes in the plane can be made this way, as a cross-section of the pyramid?

   List some plane shapes that you think *can* occur as a cross-section of the pyramid.

   List some plane shapes that you think *cannot* occur as a cross-section of the pyramid.

2. If available, use modeling clay to make a pyramid. Slice straight through the pyramid with dental floss, as if you were slicing the pyramid with a plane. Observe the cross-section that you create this way (i.e., the plane shape where the pyramid was cut). Put the pyramid back together and slice it in a different way.

   List the cross-sections you found by slicing the pyramid in various ways.

3. Think more about which plane shapes can and cannot occur as a cross-section of the pyramid. For each of the following shapes, either explain how the shape can occur as a cross-section of the pyramid or explain why it cannot occur: a trapezoid, a pentagon, a hexagon.

SECTION 13.2 CLASS ACTIVITY 13-J

Cross-Sections of a Long Rectangular Prism

CCSS CCSS SMP7, 7.G.3

Materials Modeling clay and dental floss would be helpful for this activity.

Suppose you have a very long board in the shape of a rectangular prism as pictured here:

If you saw through the board with a straight cut, the place where the board was cut makes a shape in a plane (a cross-section). In this activity, consider only cuts that go through the middle of the board, *not through the ends*.

1. Is it possible to saw the board with a straight cut in such a way that the shape formed by the cut is a square? Try to visualize if this is or is not possible.

2. Is it possible to saw the board with a straight cut in such a way that the shape formed by the cut is *not* a rectangle? (A square *is* a kind of rectangle.) Try to visualize if this is or is not possible. If the answer is yes, what kind of shape other than a rectangle can you get?

3. Use modeling dough to make a model of a long board. Make a straight slice through your model with dental floss. Describe the plane shape that the cut made. Now restore your model of the board to its original shape and make a different slice through your model. Keep trying different ways to slice your model. Describe all the different shapes you get where the model was cut.

4. Are there any plane shapes that can't arise from slicing the board? Give some examples. How do you know they can't arise? In particular, are these shapes possible: a pentagon? a quadrilateral that is not a parallelogram?

SECTION 13.3   CLASS ACTIVITY 13-K 🍎

Why the Volume Formula for Prisms and Cylinders Makes Sense

CCSS   CCSS SMP3, SMP7, 5.MD.5a

Materials Cubic-inch blocks would be helpful for parts 1, 2, and 3. You might like to have open-top boxes of the dimensions shown in part 2 or other open-top boxes. To demonstrate shearing in part 5, you might like to have a stack of index cards.

1. What does it mean for the volume of a solid object to be 12 cubic inches? How could you illustrate that meaning?

2. Suppose that students in a fifth grade class have learned what volume means, but have not yet learned about volume formulas. The students have some open-top boxes in the shape of rectangular prisms. The students also have a set of about 100 cubic inch blocks.

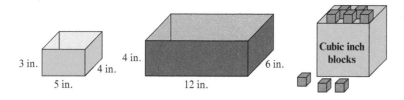

a. What is the simplest and most direct way for the students to determine the volume of a box that is 5 inches long, 4 inches wide, and 3 inches tall?

b. The students have a box that is 12 inches long, 6 inches wide, and 4 inches tall. When the students try to determine the volume of the box by filling it with blocks, they find they don't have enough blocks.

The students have learned about multiplication as equal groups. How could the students determine the volume of the box by reasoning about equal groups? Try to find several different ways they might reason!

3. Build prisms of the indicated heights on top of the bases shown below.

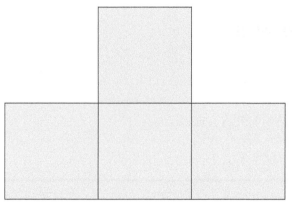

Build a 3-inch-tall prism that has this base.

Number of layers: _____

Number of cubes in each layer: _____

Expression for total number of cubes: _____

Build a 4-inch-tall prism that has this base.

Number of layers: _____

Number of cubes in each layer: _____

Expression for total number of cubes: _____

Then explain why the following volume formula for right prisms and cylinders makes sense:

$$\text{volume} = (\text{number of layers}) \cdot (\text{number of cubes in each layer})$$

4. Use the formula in part 3 to obtain this volume formula for right prisms and cylinders:

$$\text{volume} = (\text{height}) \cdot (\text{area of the base})$$

In this formula, the height is a *length* and the area of the base is an *area*. How are this length and area *linked to yet different from* the number of layers and the number of cubes in each layer in the formula in part 3?

5. Use the result of part 4 and Cavalieri's principle to explain why the formula

$$\text{volume} = (\text{height}) \cdot (\text{area of base})$$

gives the correct volume for an *oblique* prism or cylinder. Explain why the height should be measured perpendicular to the bases, and not "on the slant."

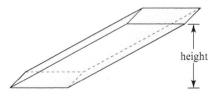

height

SECTION 13.3 CLASS ACTIVITY 13-L

Comparing the Volume of a Pyramid with the Volume of a Rectangular Prism

CCSS CCSS SMP2, 8.G.9

Materials You will need Downloads 13-12 and 13-13 at bit.ly/2SWWFUX, scissors, tape, and dry beans or rice for this activity.

1. Cut out, fold, and tape the patterns on the Downloads to make an open rectangular prism and an open pyramid with a square base.

2. Verify that the prism and the pyramid have bases of the same area and have equal heights.

   Just by looking at your shapes, make a guess: How do you think the volume of the pyramid compares with the volume of the prism?

3. Now fill the pyramid with beans, and pour the beans into the prism. (Try to keep the faces of the pyramid and prism flat while they are filled.) Keep filling and pouring until the prism is full. Based on your results, fill in the blanks in the equations that follow:

   $$\text{volume of prism} = \underline{\hspace{1cm}} \cdot \text{volume of pyramid}$$
   $$\text{volume of pyramid} = \underline{\hspace{1cm}} \cdot \text{volume of prism}$$

SECTION 13.3 **CLASS ACTIVITY 13-M**

The $\frac{1}{3}$ in the Volume Formula for Pyramids and Cones

CCSS CCSS SMP2, 8.G.9

Materials You will need Download 13-14 or 13-15 at bit.ly/2SWWFUX, scissors and tape for this activity.

This class activity will help you see where the $\frac{1}{3}$ in the volume formula for prisms comes from.

1. Cut out three of the four patterns on Download 13-14. The fourth pattern is a spare. Fold these patterns along the *undashed* line segments, and glue or tape them to make three *oblique* pyramids. Make sure the dashed lines appear on the outside of each oblique pyramid.

   If you are using Download 13-15, cut out all the pieces and tape them together to make three oblique pyramids.

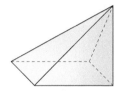

2. Fit the three oblique pyramids together to make a familiar shape. What shape is it? What is the volume of the shape formed from the three oblique pyramids? Therefore, what is the volume of one of the oblique pyramids?

3. Use your answers to parts 2 and 3 and Cavalieri's principle to explain why a *right* pyramid that is 1 unit high and has a 1-unit-by-1-unit square base has volume $\frac{1}{3}$ cubic units. (The dashed lines on the model oblique pyramids are meant to help you see how to shear the oblique pyramid. Imagine that the oblique pyramids are made out of a stack of small pieces of paper, where each dashed line going around the oblique pyramid represents a piece of paper.)

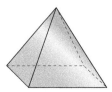

SECTION 13.3   CLASS ACTIVITY 13-N

Volume Problem Solving

CCSS CCSS SMP1, 7.G.6, 8.G.9

1. The lateral part of a cone (not including the base) is made from a half-circle of radius 10 cm. Determine the volume of the cone. Explain your reasoning.

2. Consider the block shown below.

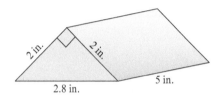

a. Which of the following can be applied to determine the volume of the block: the pyramid volume formula, the prism volume formula, both, or neither? Explain, and determine the volume of the block with a formula if it is possible to do so.

b. Determine the volume of the block in another way and explain your reasoning.

3. Determine the volume of the staircase shown below in two different ways.

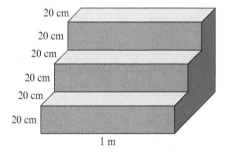

4. Determine the volume of a pyramid that has a 6-cm-by-6-cm square base and four faces that are equilateral triangles. Explain your reasoning.

SECTION 13.3 **CLASS ACTIVITY 13-O**

Deriving the Volume of a Sphere

CCSS CCSS SMP3

Half-sphere of radius r:

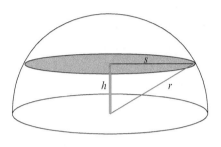

Cylinder of radius r and height r with a cone of radius r and height r removed:

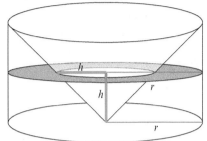

Cross-section at height h:

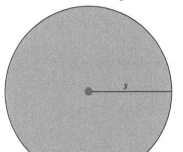

Cross-section at height h:

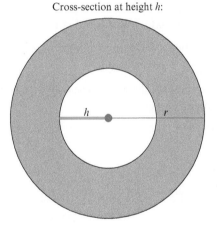

1. Explain why the colored cross-sections at height h of the half-sphere and of the part of the cylinder outside the cone have the same area.

2. Find the volume of the part of the cylinder that is outside the cone. By Cavalieri's principle and part 1, this volume is the same as the volume of the hemisphere. Use this to verify the volume formula for a sphere of radius r units.

CLASS ACTIVITY 13-P

Volume versus Surface Area and Height

CCSS CCSS SMP1, SMP2

Materials You will need several blank pieces of paper or graph paper, such as Download G-4 or G-5 at bit.ly/2SWWFUX, a ruler, scissors, and tape. Cubic inch or cubic centimeter blocks would be helpful for part 2.

1. Young children sometimes think that taller containers necessarily hold more than shorter ones. Make or describe two open-top boxes such that the taller box has a smaller volume than the shorter box.

2. Students sometimes get confused between the volume and surface area of a solid shape and about how to calculate volume and surface area.

    a. Discuss the distinction between volume and surface area. If blocks are available, use them to help you discuss the distinction.

    b. Discuss the differences and similarities in the way we calculate the volume and surface area of a rectangular prism.

3. *A cylinder volume contest:* Each team starts with a single rectangular piece of paper that is A cm by $2A$ cm, so it is twice as long as it is wide. The team cuts their paper apart and tapes it together to make the lateral surface of a cylinder (not including the bases). Each team must write a formula in terms of A for the volume of their cylinder. The team that makes the cylinder of largest volume wins the contest.

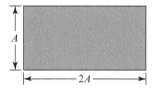

    a. Would it be possible to make a cylinder of even larger volume than the winning team's cylinder?

    b. What is the area of the lateral surface of the winning cylinder (not including the bases)? How does it compare to the area of the lateral surface of the other teams' cylinders?

    c. What is the height of the winning cylinder? How does it compare to the height of other teams' cylinders?

SECTION 13.4 CLASS ACTIVITY 13-Q

Underwater Volume Problems

CCSS CCSS SMP1, 6.G.2

1. A container can hold 2 liters. Initially, the container is $\frac{1}{2}$ full of water. When an object is placed in the container, the object sinks to the bottom, and the container becomes $\frac{2}{3}$ full. What is the volume of the object in cubic centimeters? Explain.

2. A tank in the shape of a rectangular prism is 50 cm tall, 80 cm long, and 30 cm wide. First, some rocks are placed at the bottom of the tank. Then 80 liters of water are poured into the tank. At that point, the tank is $\frac{3}{4}$ full. What is the total volume of the rocks in cubic meters? Explain.

3. A fish tank in the shape of a rectangular prism is 1 m long and 30 cm wide. The tank is $\frac{1}{2}$ full. Then 30 liters of water are poured in, and the tank becomes $\frac{2}{3}$ full. How tall is the tank? Explain your reasoning.

SECTION 13.4   CLASS ACTIVITY 13-R

Floating versus Sinking: Archimedes's Principle

CCSS CCSS SMP2, SMP4

Materials For this activity you will need a milliliter measuring cup, water, and modeling clay that does not dissolve in water. In part 3 you will need a scale for weighing in grams. In part 4 you will need a small paper cup and several coins or other small, heavy objects that fit inside the cup.

1. Pour water into your measuring cup and note the volume of water in milliliters. Form your clay into a "boat" that will float. By how much does the water level rise when you float your clay boat? Does this increase in water level tell you the volume of the clay?

2. Predict what will happen to the water level when you sink your clay boat: Will the water level go up or will the water level go down? Sink your boat and see if your prediction was correct.

   Did the increase in water level in part 1 tell you the volume of the clay, or not? Explain.

3. Exactly how much water does a floating object displace? Do the following to find out:

   a. Weigh your clay boat from part 1.

   b. Weigh an amount of water that weighs as much as your clay boat.

   c. Float your clay boat in the measuring cup and record the water level.

   d. Remove your clay boat from the water and pour the water you measured in part (b) into the measuring cup. Compare the water level now to the water level in part (c). If you did everything correctly, these two water levels should be the same.

   This experiment illustrates **Archimedes's principle** that a floating object displaces an amount of water that weighs as much as the object.

4. Determine approximately how much a quarter weighs in grams by using Archimedes's principle. To do so, partly fill a measuring cup with water. Then place a small cup in the water so that the small cup floats. Observe how much the water level rises when you put several quarters in the floating cup.

SECTION 14.1   CLASS ACTIVITY 14-A

Exploring Reflections, Rotations, and Translations with Transparencies

CCSS CCSS SMP5, 8.G.1, 8.G.3

Materials You will need a transparency, transparency markers, a paperclip, and a coordinate grid, such as Download 14-1 at bit.ly/2SWWFUX.

1. Draw a trapezoid in the top right portion of your coordinate grid. Label the 4 vertices on your trapezoid A, B, C, and D. Put the transparency on top of your coordinate grid, draw your trapezoid onto the transparency, and label the corresponding vertices as A′, B′, C′, and D′. Draw the x- and y-axes on your transparency.

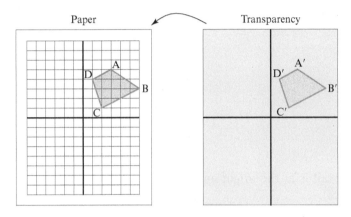

2. Perform each of the transformations in (a)–(f). Start with the transparency on top of the paper coordinate grid with the trapezoids aligned. Leave the paper fixed and move your transparency as described. In each case, answer the following: (i) is the image of a line segment a line segment of the same length? (ii) is the image of an angle an angle of the same size? (iii) are the images of parallel lines actually parallel lines?

   a. Reflect across the x-axis by twirling the transparency around the location of the x-axis so as to flip the transparency upside down.

   b. Reflect across the y-axis by twirling the transparency around the location of the y-axis so as to flip the transparency upside down.

   c. Rotate 180° around the origin by pressing the tip of an unbent paperclip onto the transparency over the origin and rotating the transparency.

d. Rotate 90° counterclockwise around the origin using a paperclip as before.

e. Rotate 90° clockwise around the origin using a paperclip as before.

f. Translate by an arrow whose tail is at the origin and whose head is at the point (_____, _____) (your choice).

3. Perform the transformations in parts (a)–(f) again. This time record the coordinates of A, B, C, D and their images. Describe how the coordinates of each point and its image are related.

SECTION 14.1 **CLASS ACTIVITY 14-B** 🍶

Reflections, Rotations, and Translations in a Coordinate Plane

CCSS CCSS SMP7, SMP8, 8.G.3

As you draw the result of translating, reflecting, or rotating a shape, it will help you to consider where the vertices of the shape will go. It may also help you to consider the relative locations of vertices within the shape.

1. a. Draw the result of translating the shaded shapes in the coordinate plane below according to the direction and the distance given by the arrow. Explain how you know where to draw your translated shapes.

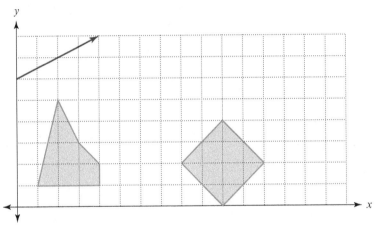

   b. For the translation in part (a), what is the image of a point (a, b)?

2. a. Draw the result of reflecting the shaded shapes in the coordinate planes below across the y-axis. Then reflect the original shapes across the x-axis. Explain how you know where to draw your reflected shapes.

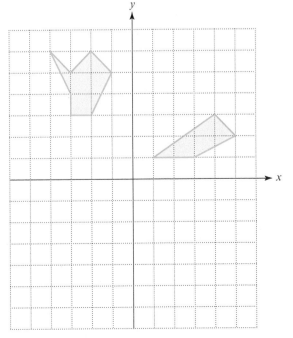
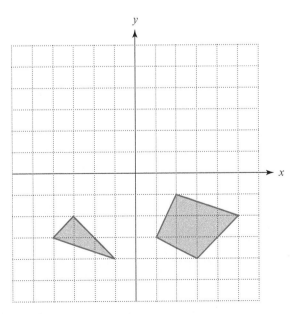

   b. What is the image of a point (a, b) when it is reflected across the y-axis?

3. a. Draw the result of rotating the shaded shapes in the coordinate planes below by 180° around the origin (where the *x*- and *y*-axes meet). Explain how you know where to draw your rotated shapes.

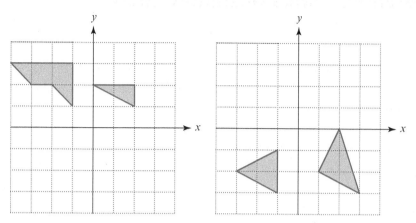

   b. What is the image of a point (a, b) when it is rotated 180° around the origin?

4. a. Draw the result of rotating the shaded shapes in the coordinate planes below by 90° counterclockwise around the origin. Explain how you know where to draw your rotated shapes.

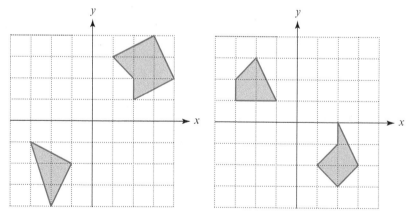

   b. What is the image of a point (a, b) when it is rotated 90° counterclockwise around the origin?

SECTION 14.1 **CLASS ACTIVITY 14-C**

Rotating and Reflecting with Geometry Tools

CCSS CCSS SMP5

Materials You will need a compass (for drawing circles), a protractor (for measuring angles), and a right angle ruler (from a geometry set—or use the corner of a piece of paper).

1. Use a compass and protractor to draw the images of:
   a. triangle ABC after a 120° counterclockwise rotation around P;

   b. triangle DEF after a 60° clockwise rotation around Q.

   Describe your methods briefly. Is there another method?

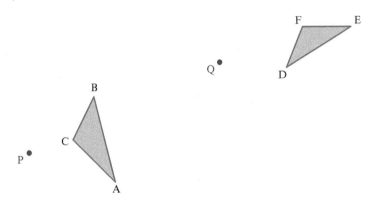

2. Use a right angle ruler (or the corner of a piece of paper) and compass to draw the images of following triangles:
   a. triangle GHI after a reflection across line ℓ

   b. triangle JKL after a reflection across line *m*

   Describe your methods briefly, explaining how to use a compass instead of a ruler to create desired distances. Is there another method?

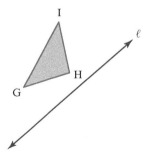

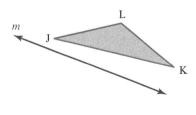

SECTION 14.1 **CLASS ACTIVITY 14-D**

Which Transformation Is It?

CCSS CCSS SMP1, SMP7

1. The figures below show two transformations taking points A, B, C, D to points A′, B′, C′, D′. What kind of transformations are they? How can you tell?

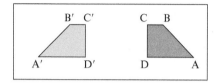
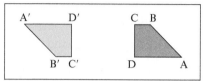

2. The figures below show two transformations taking points A, B, C to points A′, B′, C′. What kind of transformations are they? How can you tell?

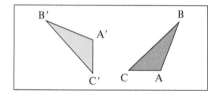
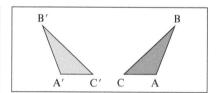

3. Does the figure show translation by the arrow that is shown? Why or why not?

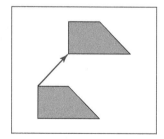

SECTION 14.2 CLASS ACTIVITY 14-E 🍎

Checking for Symmetry

CCSS CCSS SMP7, 4.G.3

Materials You will need scissors, and a paperclip or toothpick, and *two* copies of Download 14-2 at bit.ly/2SWWFUX for this activity.

Five small designs are shown below. Cut out two copies of each of these designs from two copies of the Download. For each design

- determine whether the design has reflection symmetry, and if so, what the lines of symmetry are;
- determine if the design has rotation symmetry, and if so, whether it has 2-fold, 3-fold, 4-fold, or other rotation symmetry; and
- determine whether the design has translation symmetry.

Note that true translation symmetry can occur only with designs that are infinitely long, so you may need to imagine a design continuing on forever.

To check for symmetry you may want to put one copy of a design over the other, hold them up to a light, and move one copy while you keep the other fixed.

SECTION 14.2 CLASS ACTIVITY 14-F

Traditional Quilt Designs

CCSS CCSS SMP7

1. Complete the next four traditional quilt patch designs so that each one has both a horizontal and a vertical line of symmetry.

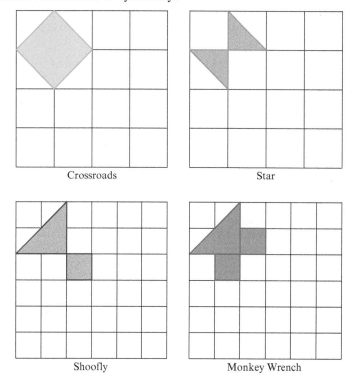

Crossroads   Star

Shoofly   Monkey Wrench

2. Complete the four traditional quilt patch designs below so that each one has 4-fold rotation symmetry.

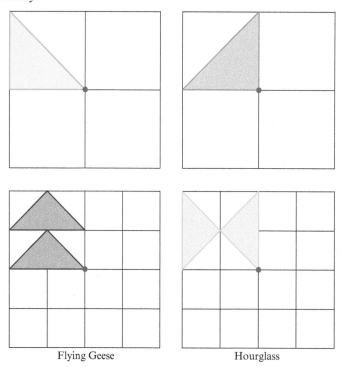

Flying Geese   Hourglass

SECTION 14.2 CLASS ACTIVITY 14-G

Creating Interlocking Symmetrical Designs

CCSS CCSS SMP1

Materials To do this activity, you must rotate and translate objects using geometry software.

1. Follow the instructions below to create your own interlocking design. Be sure to select the points, and not just the line segments, when you rotate and translate the various portions of the design. You will need some of those points to specify subsequent rotations or translations.

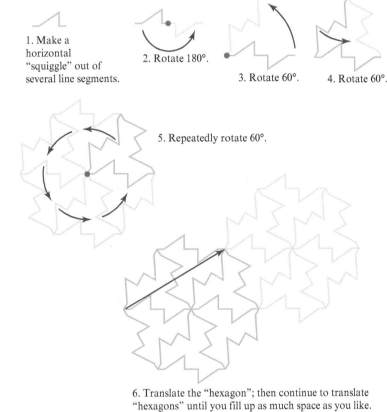

1. Make a horizontal "squiggle" out of several line segments.

2. Rotate 180°.

3. Rotate 60°.

4. Rotate 60°.

5. Repeatedly rotate 60°.

6. Translate the "hexagon"; then continue to translate "hexagons" until you fill up as much space as you like.

2. Determine the symmetries of your design. Does your design have rotation symmetry? Translation symmetry?

3. Find another way to use geometry software to create symmetrical designs.

SECTION 14.3   CLASS ACTIVITY 14-H

Motivating a Definition of Congruence

CCSS   CCSS SMP6, 8.G.2

Materials You will need scissors and Download 14-5 at bit.ly/2SWWFUX (this one download is enough for four people).

Cut out one copy of Triangle 1 and one copy of Triangle 2 from the Download and share the rest of the triangles with your classmates. Put the two triangles on the desk in front of you and move them on the desk until you can determine if they are identical copies of each other. You may also flip the triangles upside down.

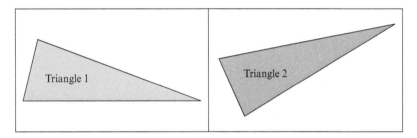

SECTION 14.3 CLASS ACTIVITY 14-I 🍎

Triangles and Quadrilaterals of Specified Side Lengths

CCSS CCSS SMP7

Materials: You will need scissors, several straws, and some string for this activity.

1. Cut a 3-inch, a 4-inch, and a 5-inch piece of straw, and thread all three straw pieces onto a piece of string. Tie a knot so as to form a triangle from the three pieces of straw.

2. Now cut two 3-inch pieces of straw and two 4-inch pieces of straw, and thread all four straw pieces onto another piece of string in the following order: 3-inch, 4-inch, 3-inch, 4-inch. Tie a knot so as to form a quadrilateral from the four pieces of straw.

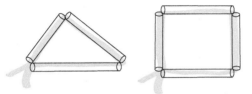

3. Compare your straw triangle and your straw quadrilateral. What is an obvious difference between them (other than the fact that the triangle is made of three pieces and the quadrilateral is made of four)?

4. When you made your triangle, if you had strung your three pieces of straw in a different order, would your triangle be different or not?

SECTION 14.3 CLASS ACTIVITY 14-J 🍎

What Information Specifies a Triangle?

CCSS CCSS SMP5, 7.G.2

Materials You will need a compass, protractor, ruler, and paper for this activity.

For each of the following, try to draw a triangle that has vertices A, B, and C and has the given specifications. Is there such a triangle? If so, think about whether any other such triangle will necessarily be congruent to yours or not. Then compare your triangle to a neighbor's. Are they congruent or not?

Triangle 1 Three side lengths are given:

From A to B is 6 cm.

From B to C is 7 cm.

From C to A is 8 cm.

Triangle 2 Three side lengths are given:

From A to B is 3 cm.

From B to C is 4 cm.

From C to A is 8 cm.

Triangle 3 Two side lengths and the angle between them are given:

From A to B is 5 cm.

The angle at A is 40°.

From A to C is 7 cm.

Triangle 4 A side length and the angle at both ends are given:

The angle at A is 30°.

From A to B is 8 cm.

The angle at B is 45°.

Triangle 5 Two side lengths and an angle that is not between them are given:

The angle at A is 20°.

From A to B is 8 cm.

From B to C is 4 cm.

Triangle 6 All three angles are given:

The angle at A is 20°.

The angle at B is 70°.

The angle at C is 90°.

SECTION 14.3   **CLASS ACTIVITY 14-K**

Why Do Isosceles Triangles and Rhombuses Decompose into Congruent Right Triangles?

CCSS CCSS SMP3

1. Given that in triangle ABC, sides AB and AC have the same length, explain why the triangle decomposes into two congruent right triangles. To start your explanation, assume that we can do either of the following:

   • Add a line segment from A to the midpoint M of side BC (so that BM and MC are the same length).

   • Add a line segment from A to side BC that divides the angle at A in half.

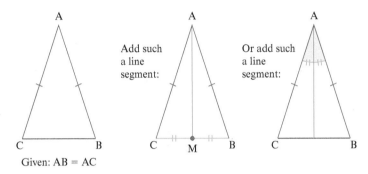

Given: AB = AC

2. What is wrong with this proposed explanation for part 1:

   > Add a line segment from A to the midpoint M of side BC that is perpendicular to side BC and divides the angle at A in half. Then because angles BAM and CAM are the same size and angles BMA and CMA are both right angles and therefore the same, it follows that angles MBA and MCA are also the same size (because the sum of the angles in a triangle is 180°). So all angles and all side lengths are the same in triangles BAM and CAM, and so those triangles are congruent right triangles.

3. Use the result of part 1 to explain why a rhombus decomposes into four congruent right triangles.

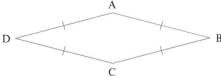

SECTION 14.3 CLASS ACTIVITY 14-L

Sewing Boxes and Congruence

CCSS CCSS SMP3, SMP4

Sewing boxes and tool boxes are sometimes constructed as shown in the side view below so that AD and BC are the same length and AB and CD are the same length. Explain why sides AB and CD are guaranteed to remain parallel (which ensures that the contents of the top drawer won't spill out when opening the box).

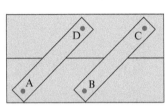

Side view of a closed sewing box.

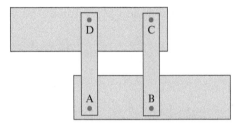

Side view of an open sewing box.

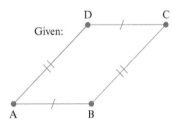

Given:

Why are sides AB and CD guaranteed to be parallel?

SECTION 14.3 **CLASS ACTIVITY 14-M**

What Was the Robot's Path?

CCSS CCSS SMP1, SMP3, SMP4

A robot moves along a (flat) factory floor as follows:

1. Starting at point A, the robot moves straight to point B.
2. At point B the robot turns 110° counterclockwise.
3. The robot moves straight to point C.
4. At point C the robot turns 70° counterclockwise.
5. The robot moves straight to point D.
6. At point D the robot turns 110° counterclockwise.
7. The robot moves straight and stops when it gets back to point A.

What can you say about the distances that the robot traveled? Must any of them be related? Explain!

SECTION 14.4 **CLASS ACTIVITY 14-N**

How Are Constructions Related to Properties of Rhombuses?

CCSS CCSS SMP7

1. Below you see the steps of a straightedge and compass construction. Draw the rhombus that arises naturally from this construction and that has A and B as vertices. Use the definition of rhombus to explain *why* the quadrilateral that you identify as a rhombus really must be a rhombus.

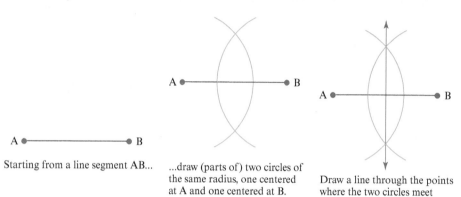

Starting from a line segment AB...

...draw (parts of) two circles of the same radius, one centered at A and one centered at B.

Draw a line through the points where the two circles meet

2. Which special properties of rhombuses explain why the line produced in the construction in part 1 really is perpendicular to the line segment AB and divides the line segment AB in half? Explain.

3. Below you see the steps of a straightedge and compass construction. Draw the rhombus that arises naturally from this construction. Use the definition of rhombus to explain *why* the quadrilateral that you identify as a rhombus really must be a rhombus.

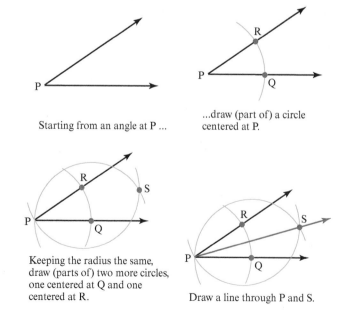

Starting from an angle at P ...

...draw (part of) a circle centered at P.

Keeping the radius the same, draw (parts of) two more circles, one centered at Q and one centered at R.

Draw a line through P and S.

4. Which special properties of rhombuses explain why the construction divides the angle at P in half? Explain.

SECTION 14.4 CLASS ACTIVITY 14-O

Construct Parallel and Perpendicular Lines by Constructing Rhombuses

CCSS CCSS SMP1, SMP5

Materials You will need a straightedge (or ruler) and a compass.

1. Construct a line that is parallel to the given line and passes through the point P by constructing a suitable rhombus using a straightedge and compass. Briefly describe your method and explain what property of rhombuses you are using.

P

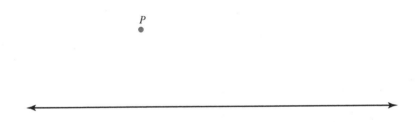

2. Construct a line that is perpendicular to the given line and passes through the point Q by constructing a suitable rhombus using a straightedge and compass. Briefly describe your method and explain what property of rhombuses you are using.

Q

SECTION 14.4   CLASS ACTIVITY 14-P

How Can We Construct Shapes with a Straightedge and Compass?

CCSS   CCSS SMP1, SMP5

Materials   You will need a straightedge (or ruler) and a compass.

1. Using a straightedge and compass, construct a line that is perpendicular to the line segment AB shown in the figure below and that passes through point A (*not* through the midpoint of line segment AB!). *Hint:* First extend the line segment AB.

    Then use a straightedge and compass to construct a square that has line segment AB as one side.

A ●————————————● B

2. Using only a straightedge and compass (no protractor for measuring angles!), construct a triangle whose angles are 45°, 45°, 90°. Explain why your method produces the desired angles.

3. Using only a straightedge and compass (no protractor for measuring angles!), construct a triangle whose angles are 30°, 60°, 90° by first constructing an equilateral triangle. Explain why your method produces the desired angles.

4. Using a straightedge and compass, construct an octagon whose vertices all lie on the circle in the next figure and whose sides all have the same length. Explain your method.

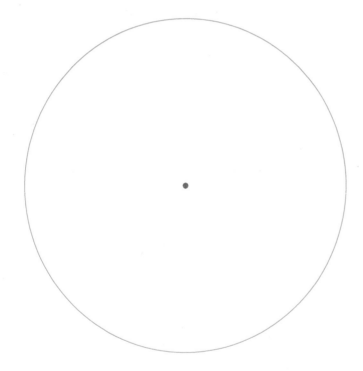

SECTION 14.5 | CLASS ACTIVITY 14-Q

Mathematical Similarity versus Similarity in Everyday Language

CCSS CCSS SMP6

In mathematics, the terms *similar* and *similarity* have a much more specific meaning than they do in everyday language.

Examine the shapes below. In everyday language, we might say that all the shapes are similar. But mathematically, only the top two shapes on the right are similar to the original shape. How is the relationship of those two shapes to the original shape different from the relationship of the other two shapes to the original shape?

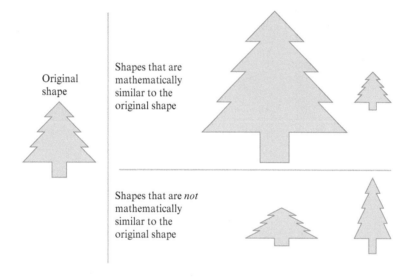

Original shape

Shapes that are mathematically similar to the original shape

Shapes that are *not* mathematically similar to the original shape

CLASS ACTIVITY 14-R

What Ways Can We Find to Solve Similarity Problems?

CCSS CCSS SMP1, 7.G.1

You have a poster that is 2 feet wide and 4 feet long. The poster has a simple design on it that you would like to scale up and draw onto a larger poster. The larger poster is to be 6 feet wide. How long should the poster be?

Find as many different ways as you can to solve this poster-scaling problem. In each case, explain your reasoning.

SECTION 14.5 **CLASS ACTIVITY 14-S** 🏺

Reasoning with the Scale Factor and Internal Factor Methods

CCSS CCSS SMP1, 7.G.1

1. Suppose that you have a postcard with an attractive picture on it and that you would like to scale up this picture and draw it onto paper that you can cut from a roll. The roll of paper is 20 inches wide, and you can cut the paper to virtually any length. If the postcard picture is 4 inches wide and 6 inches long, then how long should you cut the 20-inch-wide paper? Assume that the 4-inch side will become 20 inches long.

   Use two different methods to solve the postcard problem: the *scale factor* method and the *internal factor* method, and link each to a proportion you set up. In each case, explain why the method makes sense in as concrete a way as you can.

2. Decide whether the following problem is easier to solve with the scale factor method or with the internal factor method. Explain your answer.

   A stuffed-animal company wants to produce an enlarged version of a popular stuffed bunny. The original bunny is 6 inches wide and 11 inches tall. The enlarged bunny is to be 33 inches tall. How wide should the enlarged bunny be?

3. Decide whether the following problem is easier to solve with the scale factor method or with the internal factor method. Explain your answer.

   A toy company wants to produce a scale model of a car. The actual car is 6 feet wide and 12 feet long. The scale model of the car is to be $2\frac{1}{2}$ inches wide. How long should the scale model of the car be?

SECTION 14.5 **CLASS ACTIVITY 14-T**

Critique Reasoning about Similarity

CCSS CCSS SMP3, 7.G.1

Materials For part 3 you will need scissors and centimeter graph paper, such as Download G-5 at bit.ly/2SWWFUX.

1. *Problem:* The picture on a poster that is 4 feet wide and 6 feet long is to be scaled down and drawn onto a small poster that is 1 foot wide. How long should the small poster be?

   Johnny solves the problem this way:

   > One foot is 3 feet less than 4 feet, so the length of the small poster should also be 3 feet less than the length of the big poster. This means the small poster should be $6 - 3 = 3$ feet long.

   Is Johnny's reasoning valid? Why or why not?

2. On graph paper, plot the widths and lengths of posters that are similar to a 4-foot-wide, 6-foot-long one. What do you notice about the points? Now plot Johnny's proposed width and length from part 1. What do you notice?

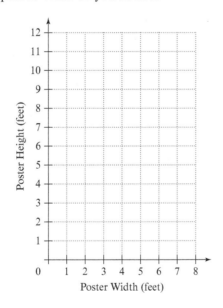

3. Using the Download, cut out the following rectangles: a 4-cm-by-6-cm rectangle, several rectangles that are similar to it, and another rectangle that is not similar to it (label this one clearly to distinguish it from the others).

   Now try this "sighting" experiment. Hold a smaller rectangle in front of you and a larger similar one behind it. Then close one eye and adjust the two rectangles until they align exactly in your line of sight in both width and length. Try to do the same with a pair of rectangles that are not similar and note the difference.

SECTION 14.6 **CLASS ACTIVITY 14-U**

Dilations versus Other Transformations: Lengths, Angles, and Parallel Lines

CCSS CCSS SMP7

Materials Optionally, you might like to use Downloads 14-6 and 14-7 at bit.ly/2SWWFUX for this activity.

Examine the results of three different transformations below and create your own examples on the next page and on the Downloads. Then discuss what the different transformations do to the angles, lines, and lengths. Do angles remain the same size? Is a transformed line parallel to the original? Do parallel lines remain parallel? How are lengths in the original related to lengths in the transformed figure? How are the *dilations* different from the other transformations?

Original

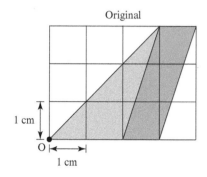

After Transformation 1 (horizontal scaling only)

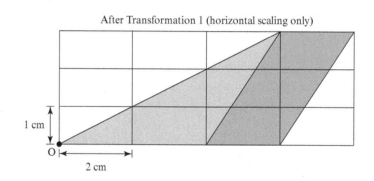

After Transformation 2
(vertical scaling only)

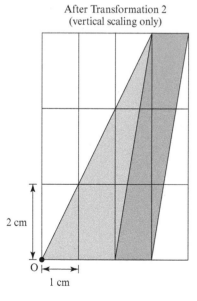

After Transformation 3, a dilation centered at O with scale factor 2

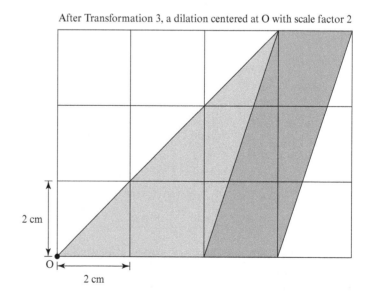

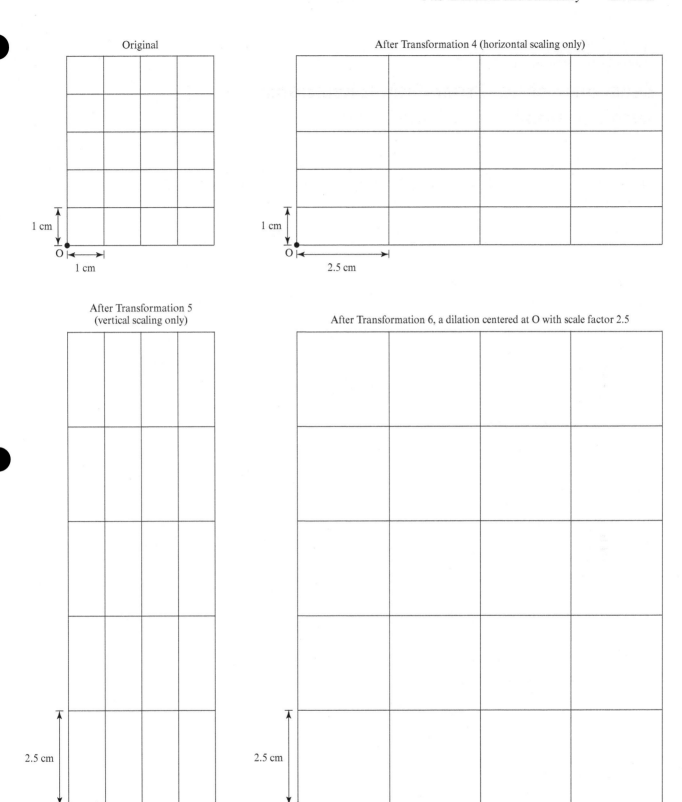

Original

After Transformation 4 (horizontal scaling only)

After Transformation 5
(vertical scaling only)

After Transformation 6, a dilation centered at O with scale factor 2.5

SECTION 14.6 **CLASS ACTIVITY 14-V**

Reasoning about Proportional Relationships with Dilations

CCSS CCSS SMP3, SMP7, SMP8

1. After applying a dilation centered at O, the figure below becomes 17 cm wide and *H* cm tall.

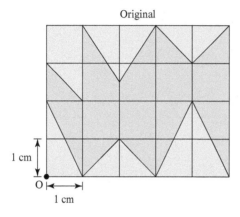

Original

1 cm

O

1 cm

 a. What is the new spacing between the grid lines? Explain!

 b. How tall does the figure become? Use our definition of multiplication (and division) to write different equations that relate *H* to other quantities. Interpret the parts of your equations in terms of the figure and the grid lines.

2. After applying a dilation centered at O, the figure above becomes *X* cm wide and *Y* cm tall.
 a. What is the new spacing between the grid lines? Explain!

 b. How are *X* and *Y* related? Use our definition of multiplication (and division) to write different equations relating *X* and *Y*. Interpret the parts of your equations in terms of the figure and the grid lines.

How Can We Measure Distances by "Sighting"?

CCSS CCSS SMP1, SMP4

Materials To do this activity, you will need your own ruler and one or more yardsticks and tape measures that can be shared by the class.

This activity will help you understand how the theory of similar triangles is used in finding distances by surveying.

1. Stand a yardstick on end on the edge of a chalkboard, or tape the yardstick vertically to the wall.

2. Stand back, away from the yardstick, in a location where you can see the entire yardstick. Your goal is to find your distance to the yardstick. Guess or estimate this distance before you continue.

   Record your guess of how far away the yardstick is here:

3. Hold your ruler in front of you with an outstretched arm. Make the ruler vertical, so that it is parallel to the yardstick. Close one eye, and with your open eye, "sight" from the ruler to the yardstick. Use the ruler to determine how big the yardstick appears to be from your location.

   Record the apparent size of the yardstick here:

4. With your arm still stretched out in front of you, have a classmate measure the distance from your sighting eye to the ruler.

   Record the distance from your sighting eye to the ruler here:

5. Use the theory of similar triangles to determine your distance to the yardstick. Sketch your eye, the ruler, and the yardstick, showing the relevant similar triangles. (Your sketch does not need to be to scale.) Explain why the triangles are similar.

   Is your calculated distance close to your estimated distance? If not, which one seems to be faulty, and why?

6. Now move to a new location, and find your distance to the yardstick again with the same technique.

SECTION 14.6   CLASS ACTIVITY 14-X

How Can We Use a Shadow or a Mirror to Determine the Height of a Tree?

CCSS   CCSS SMP1, SMP4

Materials This activity requires several tape measures that can be shared by the class. Part 2 requires a mirror.

How could you find the height of a tree, for example, without measuring it directly? This class activity provides two ways to do this with *similar triangles*. Go outside and find a tree or pole on level ground whose height you will determine. Before you continue, guess or estimate the height of the tree.

First Method

This method will work only if your tree or pole casts a fully visible shadow. (So you need a sunny day.)

1. Measure the length of the shadow of the tree (from the base of the tree to the shadow of the tip of the tree).

2. Measure the height of a classmate, and measure the length of that person's shadow.

3. Sketch the two similar triangles in this situation. Explain why the triangles are similar. Use your similar triangles to find the height of the tree.

   Is your calculated height fairly close to your estimated height of the tree? If not, which one do you think is faulty? Why?

Second Method

1. Put a mirror flat on horizontal ground away from the tree with the reflective side facing up. Stand back from the mirror at a location where you can look into the middle of the mirror and see the top of the tree (you will probably need to move around to find this location).

2. Record the following measurements:
   a. The distance from your feet to the middle of the mirror when you are looking into the middle of the mirror and can see the top of the tree.

   b. The distance from the ground to your eyes.

   c. The distance from the middle of the mirror to the base of the tree.

3. Sketch the two similar triangles in this situation. Explain why the triangles are similar. Use your similar triangles to find the height of the tree.

   Is your calculated height fairly close to your estimated height of the tree? If not, which one do you think is faulty? Why?

| SECTION 14.7 | CLASS ACTIVITY 14-Y 🝙 |

How Are Surface Areas and Volumes of Similar Boxes Related?

CCSS CCSS SMP2, SMP7, SMP8

Materials You will need scissors, tape, and Downloads 14-8 and 14-9 at bit.ly/2SWWFUX for this activity.

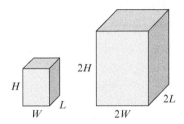

1. Cut out one box pattern from Download 14-8 (the other three are to share with classmates). Cut out the box pattern on Download 14-9. Fold these patterns to make two boxes (rectangular prisms), but leave them untaped, so that you can still unfold them. One box will have width W, length L, and height H; and one box will have width $2W$, length $2L$, and height $2H$. So the big box is twice as wide, twice as long, and twice as high as the small box.

2. By working with the patterns of the two boxes, determine how the surface area of the big box compares with the surface area of the small box. Is the surface area of the big box twice as large, three times as large, and so on, as the surface area of the small box? Look and think carefully—the answer may not be what you first think it is. Explain clearly why your answer is correct.

3. Now tape up the small box. Tape up most of the large box, but leave an opening so that you can put the small box inside it. Determine how the volume of the big box compares with the volume of the small box.

4. What if there were an even bigger box whose width, length, and height were each three times the respective width, length, and height of the small box? How would the surface area of the bigger box compare with the surface area of the small box?

   How would the volume of the bigger box compare with the volume of the small box?

5. Now imagine a variety of bigger boxes. Fill in the table below with your previous results and by extrapolating from your results.

Size of Big Box Compared with Small Box					
Length, Width, Height	2 times	3 times	5 times	2.7 times	k times
Surface area					
Volume					

SECTION 14.7 | CLASS ACTIVITY 14-Z

How Can We Determine Surface Areas and Volumes of Similar Objects?

CCSS   CCSS SMP1, SMP3, SMP4

1. If someone made a Goodyear blimp that was 1.5 times as wide, 1.5 times as long, and 1.5 times as high as the current one, how much material would it take to make the larger blimp compared with the current blimp? How much more gas would it take to fill this bigger Goodyear blimp compared with the current one (at the same pressure)?

2. An adult alligator can be 15 feet long and weigh 475 pounds. Suppose that some excavated dinosaur bones indicate that the dinosaur was 30 feet long and was shaped roughly like an alligator. How much would you expect the dinosaur to have weighed?

Alligator   Dinosaur

3. Explain why we *can't* reason in either of the following two ways to solve part 2 of this activity:

    • Each foot of the alligator weighs $475 \div 15 \approx 31.6$ pounds. So multiply that result by 30 to get the weight of the dinosaur as 950 pounds.

    • The dinosaur is twice as long as the alligator, so it should weigh twice as much, which is 950 pounds.

    What is wrong with those two ways of reasoning?

SECTION 14.7   CLASS ACTIVITY 14-AA

How Can We Prove the Pythagorean Theorem with Similarity?

CCSS CCSS SMP3, 8.G.6

This activity will help you use similar shapes to prove the Pythagorean theorem.

Remember that the Pythagorean theorem says that for any right triangle with short sides of length a and b, and hypotenuse of length c,

$$a^2 + b^2 = c^2$$

1. Given any right triangle, such as the triangle on the left below, drop the perpendicular to the hypotenuse, as shown on the right.

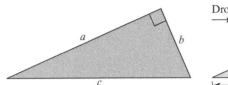 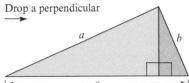

Use angles to explain why the two smaller right triangles on the right are similar to the original right triangle. (Do not use any actual measurements of angles because the proof must be general—it must work for *any* initial right triangle.)

2. Now flip each of the three right triangles of part 1 over its hypotenuse, as shown below. View each of the three right triangles as taking up a percentage of the area of the square formed on its hypotenuse. Why must each triangle take up the same percentage of its square?

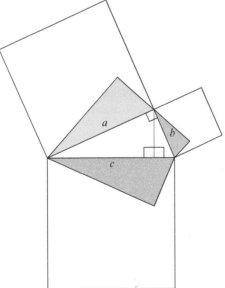

3. Let $P\%$ be the percentage of part 2. Express the areas of the three triangles in terms of $P\%$, and then explain why

$$P\% \cdot a^2 + P\% \cdot b^2 = P\% \cdot c^2$$

4. Use part 3 to explain why

$$a^2 + b^2 = c^2$$

thus proving the Pythagorean theorem.

SECTION 14.7   CLASS ACTIVITY 14-BB

Area and Volume Problem Solving

CCSS   CCSS SMP1, SMP4

1. Explain in at least two different ways why the area of the shaded region below is 3 times the area of the unshaded square inside it. (State any assumptions you make about the shapes.)

2. How does the area of the shaded region compare to the area of the unshaded region inside it? Explain. (State any assumptions you make about the shapes.)

3. A cup has a circular opening and a circular base. A cross-section of the cup and the dimensions of the cup are shown below. Determine the volume of the cup. Explain your reasoning.

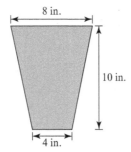

SECTION 15.1   CLASS ACTIVITY 15-A 🍎

Statistical Questions versus Other Questions

CCSS   CCSS SMP2, 6.SP.1

Materials Optionally, you may wish to use Download 15-1 at bit.ly/2SWWFUX for part 1(g).

1. Which of the following questions are *statistical* questions? What makes them statistical?
   a. How heavy is Aneeth's backpack?

   b. How heavy are the backpacks of students in this class?

   c. How much time did you spend doing homework yesterday?

   d. How much time do students at this school spend doing homework?

   e. A bag contains 50 red poker chips and 50 white poker chips. If you reach into the bag without looking and pick out 10 chips, how many red chips will you get?

   f. What is the area of a circle of radius 2 inches?

   g. When you put a circle of radius 2 inches on top of 1-inch dot paper, how many dots will be inside the circle? (You might like to try it out with the Download.)

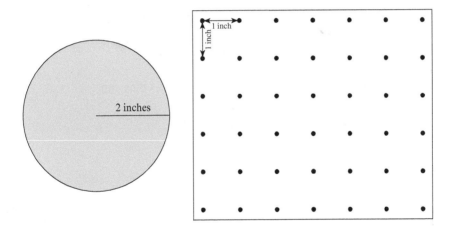

2. Write your own pair of contrasting questions, one that is statistical and one that is not. What makes one statistical and the other not?

SECTION 15.1 **CLASS ACTIVITY 15-B**

Choosing a Sample

CCSS   CCSS SMP2, 7.SP.2

A college newspaper wants to find out how students at the college would answer a specific question of importance to the student body. There are too many students for the newspaper staff to ask them all. So the staff decides to choose a sample of students to ask. For each of the following ways that the newspaper staff could select a sample, discuss whether the sample is likely to be representative of the full student body or if there are reasons why the sample may not be representative.

    a. Ask their friends.

    b. Ask as many of their classmates as they can.

    c. Stand outside the buildings their classes are in and ask as many people as they can who come by.

    d. Stand outside the student union or other common meeting area, and try to pick people who they think are representative of the students at their institution to ask the question.

    e. Generate a list of random numbers between 1 and the number of students at the college. (Many calculators can generate random numbers; random numbers can also be generated on the Internet.) Pick names out of the student directory corresponding to the random numbers (e.g., for 123, pick the 123rd name), and contact that person by phone or by e-mail.

SECTION 15.1 CLASS ACTIVITY 15-C

How Can We Use Random Samples to Draw Inferences about a Population?

CCSS CCSS SMP2, 7.SP.2

Materials Optionally, you may wish to use cups, plastic baggies, and beads, tiles, or other small objects in two different colors to simulate part 1.

1. Suppose you have a bowl filled with 250 beads, some red, the rest blue. You take a random sample of 10 beads and find that 7 are red and 3 are blue.

   Based on the random sample, how many red and blue beads should be in the full population of beads? To think about this question, imagine organizing the population of 250 beads in the following two ways:

   • **Baggies of 10 beads each:** Imagine a bunch of baggies each containing 7 red beads and 3 blue beads, just like the random sample.

   • **10 cups filled with beads:** Imagine distributing the random sample among 10 cups, so 7 cups have a red bead and 3 cups have a blue bead. Then imagine distributing all the beads among the cups.

   Explain how to reason in several different ways to infer the numbers of red and blue beads in the full population based on the random sample. In each case, support your reasoning with a math drawing and with expressions or equations involving 250, 10, 7, and 3.

2. Suppose you have a bowl filled with 650 beads, some green, the rest yellow. You take a random sample of 25 beads and find that 11 are green and 14 are yellow. Based on the random sample, how many green and yellow beads should there be? Explain how to reason in several different ways to infer those numbers. In each case, write and explain expressions for the numbers of green and yellow beads.

SECTION 15.1   CLASS ACTIVITY 15-D

Using Random Samples to Estimate Population Size by Marking (Capture–Recapture)

CCSS   CCSS SMP2, 7.SP.2

Materials   For part 1 you will need a bag filled with a large number (at least 100) of small, identical beans or other small objects that can be marked (such as small paper strips or beads that can be colored with a marker) or swapped for items of another color (such as small square tiles).

1. Pretend that the beans are fish in a lake. You will estimate the number of fish in the lake without counting them all by using a method called *capture–recapture.*

   a. Go "fishing:" Pick 60 "fish" out of your bag. Count the number of fish you caught, and label each fish with a distinctive mark (or swap it with another color). Then throw your fish back in the lake (the bag) and mix them thoroughly.

   b. Go fishing again: Randomly pick 40 fish out of your bag. Count the total number of fish you caught this time, and count how many of the fish are marked.

   c. Use your counts from parts 1 and 2 to estimate the number of fish in your bag. Explain your reasoning.

2. When Ms. Wade used the method described in part 1, she picked 30 fish at first, marked them, and put them back in the bag. Ms. Wade thoroughly mixed the fish in the bag and randomly picked out 40 fish. Of these 40 fish, 5 were marked. Explain how to reason in several different ways to determine approximately how many fish were in the bag.

SECTION 15.2 **CLASS ACTIVITY 15-E** 🏆

Critique Data Displays or Their Interpretation

CCSS CCSS SMP3

1. Ryan scooped some small plastic animals out of a tub, sorted them, and placed the animals to make a graph that looked like the following:

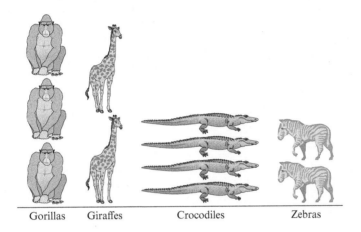

What might be a problem with Ryan's graph?

2. Critique the data display below. Would you recommend a different display?

Percentage of children ages 7 to 10 meeting dietary recommendations

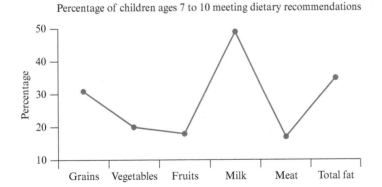

3. Critique the display below, which shows annual per capita carbon dioxide emissions in various countries, in light of Section 14.7 on how scaling affects volume.

Annual CO_2 emissions per capita in metric tons

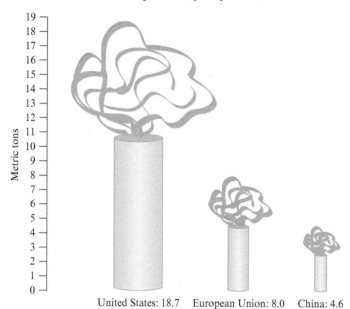

United States: 18.7   European Union: 8.0   China: 4.6

4. Students scooped dried beans out of a bag and counted the number of beans in the scoop. Each time, the number of beans in the scoop was recorded in the dot plot below.

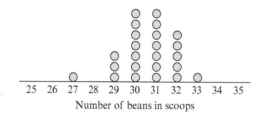

Number of beans in scoops

When a student was asked to make a list of the data displayed in the dot plot, the student responded thus:

$$1, \quad 3, \quad 7, \quad 7, \quad 5, \quad 1$$

Critique the student's response.

5. Consider the line graph below about adolescents' smoking. Based on this display, would it be correct to say that the percentage of eighth-graders who reported smoking cigarettes daily in the previous 30 days was about twice as high in 1996 as it was in 1993? Why or why not?

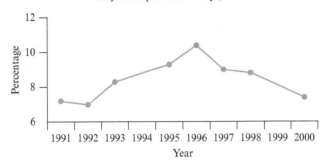

Percentage of eighth-graders who reported smoking cigarettes daily in the previous 30 days, 1991–2000

6. Consider the table below on children's eating habits.

Food	Percentage of 4–6-Year-Olds Meeting the Dietary Recommendation for a Food Group
Grains	27
Vegetables	16
Fruits	29
Saturated fat	28

Critique the pie graph below.

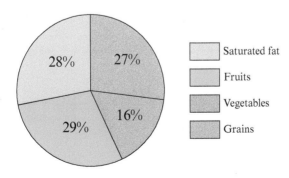

SECTION 15.2 **CLASS ACTIVITY 15-F** 🍎

Display and Ask Questions about Graphs of Random Samples

CCSS CCSS SMP4, SMP8, 7.SP.2

Materials For part 1 each pair or small group will need 20 sticky notes and either (a) a bag containing at least 30 objects that are identical except that 30% are one color and 70% are a second color or (b) use the interactive figure bit.ly/2VIjm0Q to simulate picking random samples from such a bag.

Recall that the three levels of graph comprehension discussed in the text are as follows:

- **Reading the data,** which only requires a literal reading of the graph and does not require further interpretation.

- **Reading between the data,** which requires the ability to compare quantities (e.g., greater than, tallest, smallest) or the use of other mathematical concepts and skills (e.g., addition, subtraction, multiplication, division).

- **Reading beyond the data,** which requires the student to predict or infer from the data.

1. **a.** Randomly select 10 items from the bag and write on a sticky note how many items of the first color you selected. Put the items back in the bag and repeat the process until you have 20 sticky notes.

   b. Join with another group and find a good way to organize and display your 40 sticky notes. Make a drawing of your display.

   c. Ask and answer at least two questions about your graph for each of the three graph-reading levels.

2. In your classroom you have a box with 100 small square tiles in it. The tiles are identical except that some are yellow and the rest are blue. Your students take turns picking 10 tiles out of the box without looking. Then they record the number of yellow tiles (out of the 10) they picked on a sticky note and put the tiles back in the box. The class uses the sticky notes to make a dot plot on the chalkboard. It winds up looking like the one shown below (where each dot represents a sticky note).

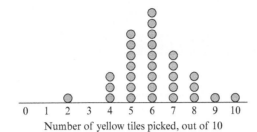

Number of yellow tiles picked, out of 10

a. Write two "read the data" questions for the dot plot. Answer your questions.

b. Write two "read between the data" questions for the dot plot. Answer your questions.

c. Write two "read beyond the data" questions for the dot plot. Answer your questions (to the extent possible).

SECTION 15.2 **CLASS ACTIVITY 15-G**

Displaying Data about Pets

CCSS CCSS SMP4, 3.MD.3

A class collected information about the pets they have at home, as shown in the table below.

Name	Pets at Home
Michelle	1 dog, 2 cats
Tyler	1 salamander, 2 snakes, 3 dogs
Antrice	1 hamster
Yoon-He	1 cat
Anne	none
Peter	2 dogs
Brandon	1 guinea pig
Brittany	1 dog, 1 cat
Orlando	none
Chelsey	2 dogs, 10 fish
Sarah	1 rabbit
Adam	none
Lauren	2 dogs
Letitia	3 cats
Jarvis	1 dog

1. Consider the following questions about pets:
   a. Are dogs the most popular pet?

   b. How many pets do most students have?

   c. How many students have more than one pet?

   d. Are most of the pets mammals?

   e. Write some other questions about pets that may be of interest to students and that could be addressed by the data that were collected.

2. Make each of the following data displays and use them to answer the questions from part 1. Observe that different graphs will be helpful for answering different questions.
   a. A display that shows how many students have 0 pets, 1 pet, 2 pets, 3 pets, and so on

   b. A display that shows how the *students* in the class fall into categories depending on what kind of pet they have

   c. Another display like the one in part (b), except pick the categories in a different way this time

   d. A display that shows how the *pets* of students in the class fall into categories

SECTION 15.2 **CLASS ACTIVITY 15-H**

Investigating Small Bags of Candies

CCSS CCSS SMP4, 2.MD.10, 3.MD.3

Materials For this activity, each person, pair, or small group in the class needs a small bag of multicolored candies. All bags should be of the same size and consist of the same type of candy. Bags should not be opened until after completion of the first part of this activity.

1. Do not open your bag of candy yet! Write a list of questions that the class as a whole could investigate by gathering and displaying data about the candies.

2. Open your bag of candy (but do not eat it yet!) and display data about your candies in two significantly different ways. For each display, write and answer questions at the three different graph-reading levels.

3. Together with the whole class, collect and display data about the bags of candies in order to answer some of the questions the class posed in part 1.

SECTION 15.2   CLASS ACTIVITY 15-I

The Length of a Pendulum and the Time It Takes to Swing

CCSS   CCSS SMP4, 5.G.2, 8.SP.1, 8.SP.2

A fifth-grader's science fair project[1] investigated the relationship between the length of a pendulum and the time it takes the pendulum to swing back and forth. The student made a pendulum by tying a heavy washer to a string and attaching the string to the top of a triangular frame, as shown in the diagram below.

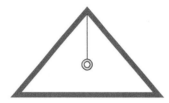

The length of the string could be varied. The table and scatterplot below show how long it took the pendulum to swing back and forth 10 times for various lengths of the string. (Several measurements were taken and averaged.)

Length of String in Inches	Time of 10 Swings in Seconds
1	2.61
2	2.97
3	3.04
4	3.41
5	3.96
6	4.13
7	4.22
8	4.5
9	4.64
10	5.13
11	5.32
12	5.56
13	5.62
14	5.87

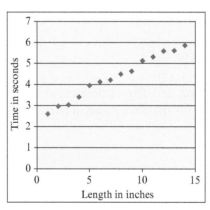

Write two or more questions about the scatterplot for each of the three graph-reading levels. Answer each question (to the extent possible).

---
[1] Thanks to Arianna Kazez for the data and information about the project.

SECTION 15.2 **CLASS ACTIVITY 15-J**

Balancing a Mobile

CCSS CCSS SMP4, 5.G.2, 8.SP.1

Materials For this activity, each person, pair, or small group in the class needs a drinking straw, string, tape, at least 7 paper clips of the same size, a ruler, and graph paper. You will use the straw, string, tape, and paper clips to make a simple mobile.

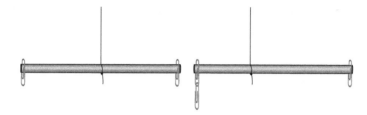

1. Tie one end of the string snugly around the straw. Tape one paper clip to each end of the straw. Hold the other end of the string so that your mobile hangs freely. Adjust the location of the string along the straw so that the straw balances horizontally. The string should now be centered on the straw, as in the picture on the left. Measure the distance on the straw from the string to each end.

2. Repeatedly add one more paper clip to one side of the straw (but not to the other side). Every time you add a paper clip, adjust the string so that the straw balances horizontally. Each time, measure the distance on the straw from the string to the end that has multiple paper clips, and record your data.

3. Make a graphical display of your data from part 2. (Use graph paper.)

4. Write and answer several questions at each of the three different graph-reading levels about your graphical display in part 3.

SECTION 15.3 CLASS ACTIVITY 15-K

The Mean as "Fair Share" by "Leveling Out"

CCSS CCSS SMP2, 5.MD.2, 6.SP.3

Materials You will need a collection of 16 small objects such as snap-together cubes or blocks for this activity. If the materials are not available, you can make drawings instead.

1. Using blocks, snap cubes, or other small objects, make groups or towers with the following number of objects in the towers:

2, 5, 4, 1

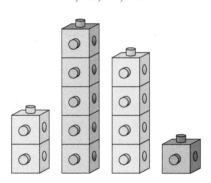

Now "level out" the block towers by sharing the blocks fairly among the 4 towers, so that all 4 towers have the same number of blocks in them. How does this common number of blocks in each of the 4 towers compare with the mean of the list 2, 5, 4, 1?

2. For each list below, make block towers. Then level out the block towers and compare the common number of blocks in each tower with the mean of the list. In some cases you may have to imagine cutting your blocks into smaller pieces.

   List 1:   1,   3,   3,   2,   1
   List 2:   6,   3,   2,   5
   List 3:   2,   3,   4,   3,   4
   List 4:   2,   3,   1,   5

3. To calculate the mean of a list of numbers *numerically,* we add the numbers and divide the sum by the number of numbers in the list. So, to calculate the mean of the list 2, 5, 4, 1, we calculate

$$(2 + 5 + 4 + 1) \div 4$$

Interpret the *numerical* process for calculating a mean in terms of leveling out 4 block towers.

   • When we add the numbers, what does that correspond to with the blocks?

   • When we divide by 4, what does that correspond to with the blocks?

SECTION 15.3 CLASS ACTIVITY 15-L

Solving Problems about the Mean

CCSS CCSS SMP1, SMP4

1. Suppose you have made 3 block towers: one 3 blocks tall, one 6 blocks tall, and one 2 blocks tall. Describe some ways to make 2 more towers so that there is an average of 4 blocks in all 5 towers. Explain your reasoning.

2. If you run 3 miles every day for 5 days, how many miles will you need to run on the sixth day in order to have run an average of 4 miles per day over the 6 days? Solve this problem in two different ways, and explain your solutions.

3. The mean of 3 numbers is 37. A fourth number, 41, is included in the list. What is the mean of the 4 numbers? Explain your reasoning.

4. Explain how you can quickly calculate the average of the following list of test scores without adding the numbers:

$$81, \quad 78, \quad 79, \quad 82$$

5. If you run an average of 3 miles a day over 1 week and an average of 4 miles a day over the next 2 weeks, what is your average daily run distance over that 3-week period?

   Before you solve this problem, explain why it makes sense that your average daily run distance over the 3-week period is *not* just the average of 3 and 4—namely, 3.5. Should your average daily run distance over the 3 weeks be greater than 3.5 or less than 3.5? Explain how to answer this without a precise calculation. Now determine the exact average daily run distance over the 3-week period. Explain your solution.

SECTION 15.3   CLASS ACTIVITY 15-M

The Mean as "Balance Point"

CCSS   CCSS SMP2, 6.SP.2, 6.SP.3

1. For each of the data sets below:
    • Make a dot plot of the data on the given axis.
    • Calculate the mean of the data.
    • Verify that the mean agrees with the location of the given fulcrum.
    • Answer this question: Does the dot plot look like it would balance at the fulcrum (assuming the axis on which the data is plotted is weightless)?

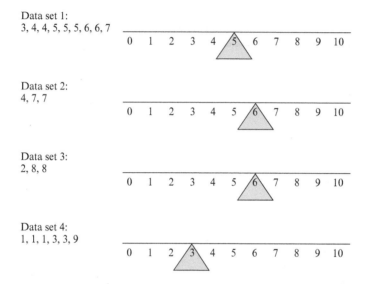

Data set 1:
3, 4, 4, 5, 5, 5, 6, 6, 7

Data set 2:
4, 7, 7

Data set 3:
2, 8, 8

Data set 4:
1, 1, 1, 3, 3, 9

2. For each of the dot plots below, guess the approximate location of the mean by thinking about where the balance point for the data would be. Then check how close your guess was by calculating the mean.

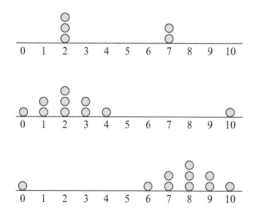

3. Previously, we discussed how to interpret the mean as a fair share, in which we "level out" the data, to make all the data equal. How is this "fair share" view of the mean related to the "balance point" view?

Examine and discuss how the next displays show the leveling-out process in terms of a dot plot. During the process, does the balance point of the dot plot change or not?

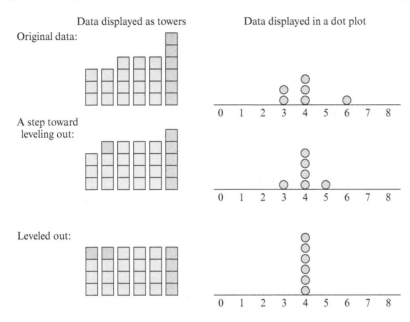

4. Interpret the "leveling out" process for determining the mean in terms of the next dot plot. As you change the data, do you change its balance point?

5. Some students are debating whether the data in the next dot plot is "leveled out" into fair shares and shows that the mean is 2. What do you think?

SECTION 15.3   CLASS ACTIVITY 15-N

Same Median, Different Mean

CCSS   CCSS SMP2, 6.SP.5d

Materials You will need 9 pennies or other small objects for this activity.

The following data set is represented on the dot plot below with pennies:

$$4, \quad 5, \quad 5, \quad 6, \quad 6, \quad 6, \quad 7, \quad 7, \quad 8$$

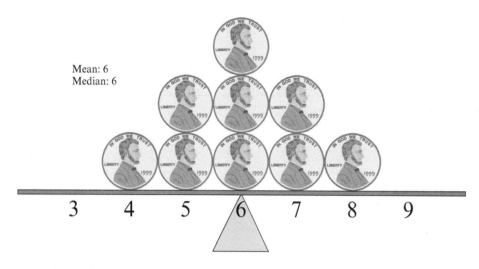

Mean: 6
Median: 6

Arrange real pennies (or other small objects) along the number line below to represent the same data set.

3   4   5   6   7   8   9

Show data sets with the same median, different means.

1. Rearrange your pennies so that they represent new data sets that still have median 6 but have means *less than* 6. To help you do this, think about the mean as the balance point. List your new data sets.

2. Rearrange your pennies so that they represent new data sets that still have median 6 but have means *greater than* 6. To help you do this, think about the mean as the balance point. List your new data sets.

SECTION 15.3   **CLASS ACTIVITY 15-O**

Can More Than Half Be Above Average?

CCSS CCSS SMP2, 6.SP.5d

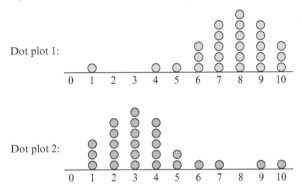

Dot plot 1:

Dot plot 2:

1. For each of the dot plots, decide which is greater: the median or the mean of the data. Explain how you can tell without calculating the mean.

2. A teacher gives a test to a class of 20 students.
   a. Is it possible that 90% of the class scores above average? If so, give an example of test scores for which this is the case. If not, explain why not.
   b. Is it possible that 90% of the class scores below average? If so, give an example of test scores for which this is the case. If not, explain why not.

3. A radio program describes a fictional town in which "all the children are above average." In what sense is it possible that all the children are above average? In what sense is it not possible that all the children are above average?

SECTION 15.3 **CLASS ACTIVITY 15-P**

Critique Reasoning about the Mean and the Median

CCSS CCSS SMP3

1. When Eddie was asked to determine the mean of the data shown in the dot plot below, he calculated thus:

   $$1 + 2 + 4 + 2 + 1 = 10, \quad 10 \div 5 = 2$$

   Eddie concluded that the mean is 2. Critique Eddie's reasoning.

2. Brittany said that the mean of the test data in the dot plot below is 3 because "all the students got the same score; all the scores are level at 3." Discuss!

   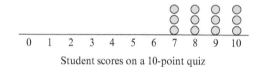

   Student scores on a 10-point quiz

3. Critique the work of Student 1, Student 2, and Student 3 below.

   Student 1:   5, 5, 6, 6, 6, 3, 4, 5, 4, 5, 5, 6, 4, 7, 5, 5, 5, 4, 6, 5

   Median: 5

   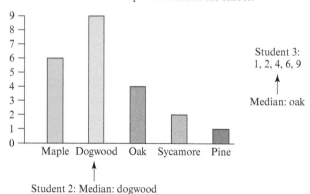

   What kind of tree should we plant in front of the school?

   Student 3:
   1, 2, 4, 6, 9

   Median: oak

   Student 2: Median: dogwood

SECTION 15.4 CLASS ACTIVITY 15-Q

What Does the Shape of a Data Distribution Tell about the Data?

CCSS CCSS SMP4, 6.SP.2

Examine histograms 1, 2, and 3 below and observe the different shapes these distributions take.

- The shape of histogram 1 is called *skewed to the right* because the histogram has a long tail extending to the right.

- The shape of histogram 2 is called *bimodal* because the histogram has two peaks.

- The shape of histogram 3 is called *symmetric* because the histogram is approximately symmetrical.

Histogram 1 Distribution of household income in the United States in 2019

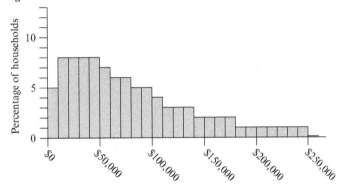

Histogram 2 Distribution of household income in hypothetical country A

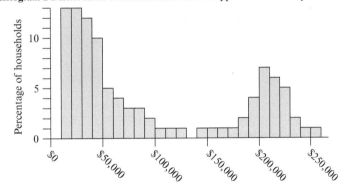

Histogram 3 Distribution of household income in hypothetical country B

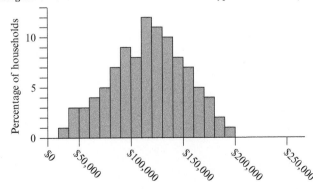

1. Write at least three questions about the graphs, including at least one question at each of the three levels of graph comprehension discussed in Section 15.2. Answer your questions.

2. Discuss what the shapes of histograms 1, 2, and 3 tell you about household income in the United States versus in the hypothetical countries A and B. What do you learn from the shape of the histograms that you wouldn't be able to tell just from the medians and means of the data?

3. Discuss how each country could use the histograms to argue that its economic situation is better than at least one of the other two countries.

SECTION 15.4   CLASS ACTIVITY 15-R

Distributions of Random Samples

CCSS   CCSS SMP4, 7.SP.2

Materials   For parts 1–4 you will need either (a) a bag filled with a large collection of small objects (at least 50) that are identical except that 40% are one color and 60% are another color, or (b) use the interactive figure bit.ly/2VIjm0Q to simulate picking random samples from such a bag.

Think of the objects as representing a group of voters. The 40% in one color represent "yes" votes and the 60% in the other color represent "no" votes.

1. You will be picking random samples of 10 from the bag and plotting the percentage of "yes" votes on a dot plot. Before you start picking these random samples, make a dot plot below to predict what your actual dot plot will look like approximately. Assume that you will plot about 30 dots.

    • Why do you think your dot plot might turn out that way?

    • How do you think the fact that 40% of the votes in the bag are "yes" votes might be reflected in the dot plot?

    • What kind of shape do you predict your dot plot will have?

Predicted _____
            0   10   20   30   40   50   60   70   80   90   100
          Percentage of "yes" votes in random samples of 10 objects

2. Now pick about 30 random samples of 10 objects from the bag. Each time you pick a random sample of 10, determine the percentage of "yes" votes and plot this percentage in a dot plot on the next page. Then return your sample to the bag.

    Compare your results with your predictions in part 1. Do you see the fact that 40% of the votes in the bag are "yes" votes reflected in the dot plot? If so, how? What kind of shape does the dot plot have?

For part 2: _____

0   10   20   30   40   50   60   70   80   90   100
Percentage of "yes" votes in random samples of 10 objects

For part 3: _____

0   10   20   30   40   50   60   70   80   90   100
Percentage of "yes" votes in random samples of 20 objects,
rounded to the nearest ten

3. Next, you will repeat part 2, but this time you will pick random samples of 20 objects from the bag, and you will round your percentages to the nearest ten. Before you start, how do you think this new dot plot will compare with your dot plot for part 2? After you are done, compare the two dot plots.

4. Your bag contains 40% "yes" votes. What range of percentages would you consider to be "pretty good predictions" of this 40%? For this range:

a. What percent of the dots in your dot plot for part 2 are "pretty good predictions"?

b. What percent of the dots in your dot plot for part 3 are "pretty good predictions"?

c. If you were to pick random samples of 25 objects from the bag, how do you think the percentage of "pretty good predictions" would compare with (a) and (b)?

5. Compare the two histograms below. The first histogram shows the percentage of "yes" votes in 200 samples of 100 taken from a population of 1,000,000 in which 40% of the population votes "yes." The second histogram shows the percentage of "yes" votes in 200 samples of 1000 taken from the same population.

   a. Compare the way the data are distributed in each of these histograms and compare these histograms with your dot plots in parts 2 and 3.

   b. How is the fact that 40% of the population votes "yes" reflected in these histograms?

   c. What do the histograms indicate about using samples of 100 versus samples of 1000 to predict the outcome of an election?

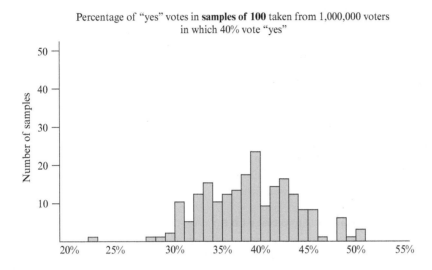

Percentage of "yes" votes in **samples of 100** taken from 1,000,000 voters
in which 40% vote "yes"

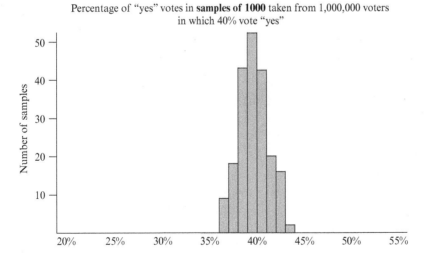

Percentage of "yes" votes in **samples of 1000** taken from 1,000,000 voters
in which 40% vote "yes"

6. What if we made a histogram like the ones above by using the same population, but by picking 200 samples of 500 (instead of 200 samples of 100 or 1000)? How do you think this histogram would compare with the ones above? What if samples of 2000 were used?

SECTION 15.4 CLASS ACTIVITY 15-S

Comparing Distributions: Mercury in Fish

CCSS   CCSS SMP4, 6.SP.2

The two histograms below display hypothetical data about amounts of mercury found in 100 samples of each of two different types of fish. Mercury levels above 1.00 parts per million are considered hazardous.

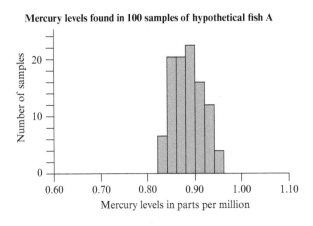

Mercury levels found in 100 samples of hypothetical fish A

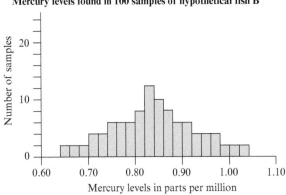

Mercury levels found in 100 samples of hypothetical fish B

1. Discuss how the two distributions compare and what this tells you about the mercury levels in the two types of fish. In your discussion, take the following into account: median or mean levels of mercury in each type of fish and the hazardous level of 1.00 parts per million.

2. In comparing the two types of fish, if you hadn't been given the histograms, would it be adequate just to have the medians or means of the amount of mercury in the samples, or is it useful to know additional information about the data?

Using Medians and Interquartile Ranges to Compare Data

CCSS CCSS SMP4, 6.SP.3, 6.SP.5

1. Determine the median, the first and third quartiles, and the interquartile range for each of the hypothetical taste-tester data shown in the following dot plots:

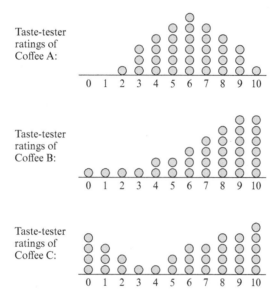

2. Discuss and compare the ratings of the three coffees using the medians and interquartile ranges from part 1. Which coffee seems to taste best? Which coffee seems to be most consistent?

SECTION 15.4   CLASS ACTIVITY 15-U

Using Box Plots to Compare Data

CCSS   CCSS SMP4, 6.SP.4

1. Make box plots for the 3 dot plots of hypothetical quiz scores in 3 classes that follow.

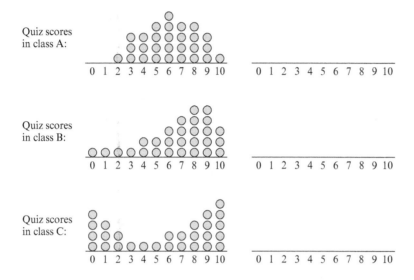

2. Suppose you had the box plots from part 1, but you didn't have the dot plots. Discuss what you could tell about how the 3 data sets are distributed. Compare how the three classes did on the quiz.

SECTION 15.4 CLASS ACTIVITY 15-V

Percentiles versus Percent Correct

CCSS CCSS SMP3

1. Determine the 75th percentile for each set of hypothetical test data in the 3 dot plots below.

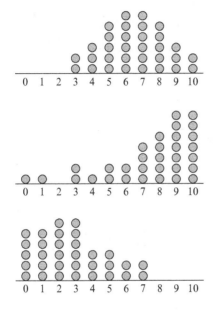

2. Discuss: On a test, is the 75th percentile the same as 75% correct?

3. Mrs. Smith makes an appointment to talk to her son Johnny's teacher. Johnny has been getting As in math, but on the standardized test he took, he was at the 80th percentile. Mrs. Smith is concerned that this means Johnny is really doing B work in math, not A work. If you were Johnny's teacher, what could you tell Mrs. Smith?

SECTION 15.4   CLASS ACTIVITY 15-W

Comparing Paper Airplanes

CCSS   CCSS SMP4, 6.SP.3, 6.SP.5c

The dot plots below show the distances, rounded to the nearest foot, that three different models of paper airplanes flew in ten trials each.

1. For each dot plot, compute the MAD by first replacing each dot with its distance from the mean as shown below the first dot plot.

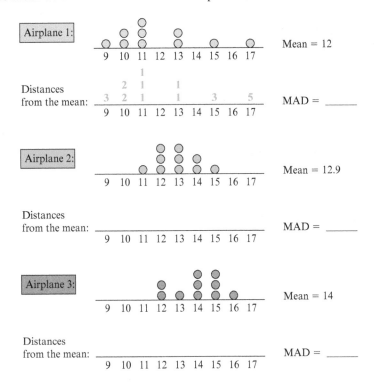

2. Use the mean and the MAD from each data set in part 1 to discuss how the three paper airplanes compare. What do the means tell you? What do the MADs tell you? How confident can you be in saying that one airplane flies farther than another?

3. Suppose there is a fourth paper airplane for which the mean is 13.9 and the MAD is 1.1. Discuss how this fourth airplane compares with airplanes 2 and 3. Discuss how confident you can be in saying that one airplane flies farther than another.

SECTION 15.4 CLASS ACTIVITY 15-X

Why Does Variability Matter When We Compare Groups?

CCSS   CCSS SMP4, 7.SP.3, 7.SP.4

Some researchers are doing an experiment to see if a treatment is effective for an illness. They randomly assign patients to two groups: a treatment group and a control group. They then determine how long it takes each patient to recover from the illness.

In this activity you will consider two hypothetical scenarios for the experiment and think about why we should take measures of center *and* of variability into account when comparing the treatment and control groups.

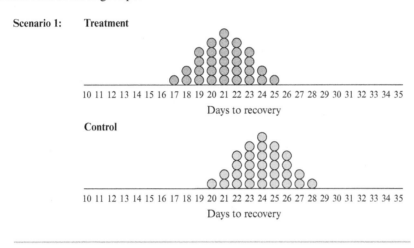

1. Compare the two hypothetical scenarios. How are they similar? How are they different?

2. If you have not done so already, discuss the following questions.

   a. How much better does the treatment group do than the control group? Can you quantify that?

   b. Does the treatment seem to have more of an effect in one scenario than in the other? How so?

3. If you have not done so already, compute the means and MADs for all four data sets (the treatment and control groups for both scenarios). For each scenario, how does the difference in means between treatment and control groups compare with the MAD? How is that related to the difference in the effect of the treatment in the two scenarios?

SECTION 15.4 CLASS ACTIVITY 15-Y

How Well Does Dot Paper Estimate Area?

CCSS CCSS SMP4, 7.SP.3, 7.SP.4

Materials You will need scissors and Downloads 15-2 and 15-3 at bit.ly/2SWWFUX for this activity.

Biologists sometimes use dot paper to estimate areas of leaves. If the dots are on a 1-inch grid, as in Download 15-2, then the number of dots the leaf covers gives an estimate for the area of the leaf in square inches.

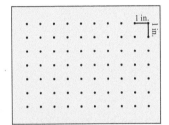
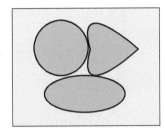

1. Cut out the circle, the oval, and the leaf-shape on Download 15-3. For each of those three shapes:
   • make a dot plot of how many dots the shape covers as the shape is placed randomly on the dot paper (count only dots that are at least half-covered by the shape);
   • compute the mean number of dots the shape covers;
   • compute the MAD of the number of dots the shape covers.

2. Use your mean numbers of dots from part 1 as estimates for the areas (in square inches) of the three shapes. Based on these estimates, which shape has the least area and which has the greatest?

3. Discuss and use the variability you found in your data in part 1.
   • Which shape exhibited the least variability? Are you surprised?
   • Use the variability to discuss how close your estimated areas from part 2 are likely to be to the actual areas and how confident you are in identifying which shape has the least area and which has the greatest.

4. Compare your results from parts 2 and 3 with this information: the circle has radius 2 inches and the oval has the same area as the circle.

SECTION 16.1 **CLASS ACTIVITY 16-A**

Probabilities with Spinners

CCSS   CCSS SMP3, 7.SP.7a

Many children's games use "spinners." You can make a simple spinner by placing the tip of a pencil through a paper clip and holding the pencil so that its tip is at the center of the circle as shown. The paper clip should spin freely around the pencil tip.

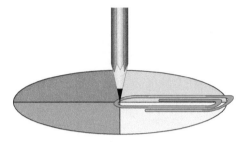

1. Compare the two spinners shown below. For which spinner is a paper clip most likely to land in a shaded region? Explain your answer.

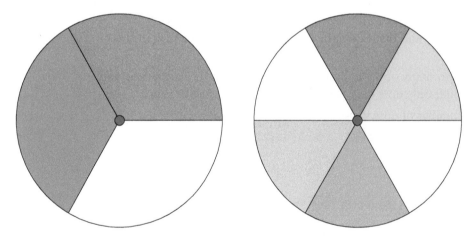

2. Compare the two spinners below. For which spinner is a paper clip most likely to land in a shaded region? Explain your answer.

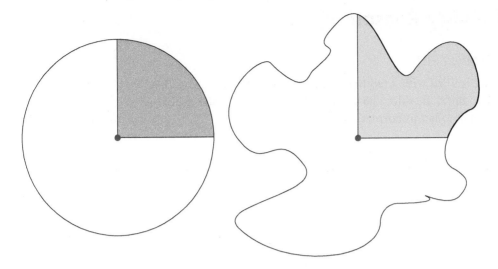

3. Draw a 4-color spinner (red, green, yellow, blue) such that
  - landing on green is twice as likely as landing on red;
  - landing on yellow is equally likely as landing on green;
  - landing on blue is more likely than landing on yellow.

Determine the probabilities of landing on each of the colors on your spinner and explain your reasoning.

Could someone else make a different spinner that has the same properties above but has different probabilities than yours?

SECTION 16.1 CLASS ACTIVITY 16-B 🍎

Critique Probability Reasoning

CCSS CCSS SMP3, 7.SP.6, 7.SP.7a

1. Kevin has a bag that is filled with 2 red balls and 1 white ball. Kevin says that because there are two different colors he could pick from the bag, the probability of picking the red ball is $\frac{1}{2}$. Is this correct? Why or why not?

2. A family math night at school features the following game. There are two opaque bags, each containing red blocks and yellow blocks. Bag 1 contains 2 red blocks and 4 yellow blocks. Bag 2 contains 4 red blocks and 16 yellow blocks. To play the game, you pick a bag and then you pick a block out of the bag without looking. You win a prize if you pick a red block. Eva thinks she should pick from bag 2 because it has more red blocks than bag 1. Is Eva more likely to pick a red block if she picks from bag 2 than from bag 1? Why or why not?

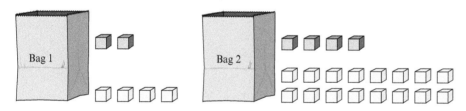

3. The probability of winning a game is $\frac{3}{1000}$. Does this mean that if you play the game 1000 times, you will win 3 times? If not, what does the probability of $\frac{3}{1000}$ stand for?

SECTION 16.1   CLASS ACTIVITY 16-C

What Do We Expect in the Long Run?

CCSS CCSS SMP4, SMP7, 7.SP.5

Consider a spinner that has 5 equal sectors, 1 red, 1 blue, and 3 yellow. When you spin the spinner, the arrow is equally likely to land in each of the 5 sectors.

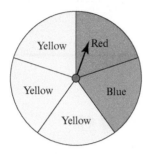

1. Imagine you could spin the spinner many times, such as (a) 500 times or (b) 2000 times, or (c) X times, where X is a large number. In each case, approximately how many times would you expect the arrow to land in red? Why?

2. Imagine you could spin the spinner many times, such as (a) 500 times or (b) 2000 times, or (c) X times, where X is a large number. In each case, approximately how many times would you expect the arrow to land in yellow? Why?

3. Approximately how many times would you expect to have to spin the spinner for the arrow to land in yellow 1500 times? Explain your reasoning.

SECTION 16.1 CLASS ACTIVITY 16-D 🍎

Empirical versus Theoretical Probability: Picking Cubes from a Bag

CCSS CCSS SMP4, 7.SP.6

Materials Each person (or small group) will need an opaque bag, 3 red cubes, 7 blue cubes, and a sticky note.

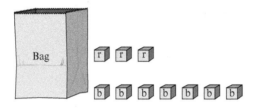

1. Put the 10 cubes in the bag, mix them up, and randomly pick a cube from the bag without looking. Record the color of the cube, and put the cube back in the bag. Repeat until you have picked 10 cubes. Record the number of red cubes you picked on your sticky note. Calculate the empirical probability of picking a red cube based on your 10 picks. Is it the same as the theoretical probability of picking red?

2. Use everyone's sticky notes from part 1 to create a class dot plot. How is the fact that there are 3 red cubes and 7 blue cubes in the bag reflected in the dot plot?

3. Use the class dot plot from part 2 to determine an empirical probability of picking a red cube from a bag containing 3 red cubes and 7 blue cubes. Compare this empirical probability with the theoretical probability of picking red.

Using Empirical Probability to Make Predictions

CCSS CCSS SMP4, 7.SP.6

A family math night at school includes the following activity. A bag is filled with 10 small counting bears that are identical except that 4 are yellow and the rest are blue. A sign next to the bag gives instructions for the activity:

> Win a prize if you guess the correct number of yellow bears in the bag! There are 10 bears in the bag. Some are yellow and the rest are blue. Here's what you do:
> - Reach into the bag, mix well, and pick out a bear.
> - Get a sticky note that is the same color as your bear, write your name and your guess on the note and add your note to the others of the same color.
> - Put your bear back in the bag, and mix well.

The sticky notes will be organized into columns of 10, so it will be easy to count up how many of each there are.

1. How will students be able to use the results of this activity to estimate the number of yellow bears in the bag?

2. What do you expect will happen as the night goes on and more and more bears are picked?

3. Discuss any additions or modifications you would like to make to the activity if you were going to use it for math night at your school.

SECTION 16.1 **SECTION 16.1** CLASS ACTIVITY 16-F

If You Flip 10 Pennies, Should Half Come Up Heads?

CCSS CCSS SMP4, 7.SP.7b

Materials You will need a bag, 10 pennies or 2-color counters, and some sticky notes for this activity.

1. Make a guess: What do you think the probability is of getting exactly 5 heads on 10 pennies when you dump the 10 pennies out of the bag?

2. Put the 10 pennies in the bag, shake them up, and dump them out. Record the number of heads on a sticky note. Repeat this for a total of 10 times, using a new sticky note each time. Out of these 10 tries, how many times did you get 5 heads? Therefore, what is the experimental probability of getting 5 heads based on your 10 trials?

3. Now work with a large group (e.g., the whole class). Collect the whole group's data on the sticky notes from part 2. Find a way to display these data so that you can see how often the whole group got 5 heads and other numbers of heads.

4. Is the probability of getting exactly 5 heads from 10 coins 50%? What does your data display from part 3 suggest?

SECTION 16.2 CLASS ACTIVITY 16-G

How Many Keys Are There?

CCSS CCSS SMP4, SMP7, 7.SP.8b

Have you ever wondered: how can millions of *different* keys be produced, even though keys are not very big? Keys are manufactured to be distinct from one another by the way they are notched. Many keys have intricate notching. For simplicity, in this activity let's consider only simple keys that are notched on one side.

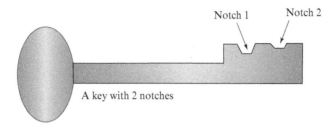

A key with 2 notches

1. Suppose a simple type of key is to be made with 2 notches, and that each notch can be one of 3 depths: deep, medium, or shallow. How many different keys can be made this way? Explain.

2. Explain how to use multiplication to solve the problem in part 1 if you haven't already.

3. Now suppose the key is to be made with 4 notches, and each notch can be one of 3 depths: deep, medium, or shallow. How many keys can be made this way? Explain.

4. Now suppose the key is to be made with 10 notches, and each notch can be one of 5 depths. How many keys can be made this way? Explain.

SECTION 16.2   CLASS ACTIVITY 16-H

Counting Outcomes: Independent versus Dependent

CCSS   CCSS SMP4, SMP7, 7.SP.8b

1. How many 3-letter security codes can be made from the 4 letters A, B, C, D? For example, BAB and ABB are two such codes, and DAC is another. Explain.

2. How many 3-letter security codes can be made from the 4 letters A, B, C, D without using a letter twice? For example, BAC and ADB are two such codes. Explain.

3. Explain how to use multiplication to solve the problems in parts 1 and 2 if you haven't already.

4. Contrast how you use multiplication to solve the problems in parts 1 and 2 and explain the distinction.

5. How many 4-letter security codes can be made from the 6 letters U, V, W, X, Y, Z? Explain.

6. How many 4-letter security codes can be made from the 6 letters U, V, W, X, Y, Z, without using a letter twice? Explain.

7. Explain how to use multiplication to solve the problems in parts 5 and 6 if you haven't already.

8. Contrast how you use multiplication to solve the problems in parts 5 and 6 and explain the distinction.

SECTION 16.3   CLASS ACTIVITY 16-I

Number Cube Rolling Game

CCSS CCSS SMP4, SMP7, 7.SP.8

Materials You will need a pair of number cubes (dice) for part 1.

Maya, James, Kaitlyn, and Juan are playing a game in which they take turns rolling a pair of number cubes. Each student has chosen a "special number" between 2 and 12, and each student receives 8 points whenever the total number of dots on the two number cubes is their special number. (They receive their points regardless of who rolled the number cubes. Their teacher picked 8 points so that the students would practice counting by 8s.)

- Maya's special number is 7.
- James's special number is 10.
- Kaitlyn's special number is 12.
- Juan's special number is 4.

The first person to get to 100 points or more wins. The students have played several times, each time using the same special numbers. They notice that Maya wins most of the time. They are wondering why.

1. Roll a pair of number cubes many times, and record the total number of dots each time. Display your data so that you can compare how many times each possible number between 2 and 12 has occurred. What do you notice?

2. Draw an array showing all possible outcomes on each number cube when a pair of number cubes are rolled. (Think of the pair as *number cube 1* and *number cube 2*.)

   a. For which outcomes is the total number of dots 7? 10? 12? 4?

   b. What is the probability of getting 7 total dots on a roll of two number cubes? What is the probability of getting 10 total dots on a roll of two number cubes? What about for 12 and 4?

   c. Is it surprising that Maya kept winning?

SECTION 16.3 **CLASS ACTIVITY 16-J** 🍎

Picking Two Marbles from a Bag of 1 Black and 3 Red Marbles

CCSS CCSS SMP4, SMP7, 7.SP.8

Materials You will need an opaque bag, 3 red marbles, and 1 black marble for this activity. Put the marbles in the bag.

If you reach in without looking and randomly pick out 2 marbles at once, what is the probability that 1 of the 2 marbles you pick is black? You will study this question in this activity.

1. Before you continue, make a guess: What do you think the probability of picking the black marble is when you randomly pick 2 marbles out of the 4 marbles (3 red, 1 black) in the bag?

2. Pick 2 marbles out of the bag at once. Repeat this many times, recording what you pick each time. What fraction of the times did you pick the black marble?

3. Now calculate the probability theoretically, using a tree diagram. For the purpose of computing the probability, think of first picking one marble, then (without putting this marble back in the bag) picking a second marble. From this point of view, draw a tree diagram that will show all possible outcomes for picking the two marbles. But draw this tree diagram in a special way, *so that all outcomes shown by your tree diagram are equally likely.*

   Hints: The first stage of the tree should show all possible outcomes for your first pick. Remember that all branches you show should be equally likely. In the second stage, the branches you draw should depend on what happened in the first stage. For instance, if the first pick was the black marble, then the second pick must be one of the three red marbles.

   a. How many total outcomes for picking 2 marbles, 1 at a time, out of the bag of 4 (3 red, 1 black) does your tree diagram show?

   b. In how many outcomes is the black marble picked (on 1 of the 2 picks)?

   c. Use your answers to parts 3 (a) and (b) and the basic principles of probability to calculate the probability of picking the black marble when you pick 2 marbles out of a bag filled with 1 black and 3 red marbles.

4. Why was it important to draw the tree diagram so that all outcomes were equally likely?

5. Here's another method for calculating the probability of picking the black marble when you pick 2 marbles out of a bag filled with 1 black and 3 red marbles:

    a. How many unordered pairs of marbles can be made from the 4 marbles in the bag?

    b. How many of those pairs of marbles in part (a) contain the black marble? (Use your common sense.)

    c. Use parts (a) and (b) and basic principles of probability to determine the probability of picking the black marble when you pick 2 marbles out of a bag containing 1 black and 3 red marbles.

6. Compare your answers to parts 3(a) and 5(a), and compare your answers to 3(b) and 5(b). How and why are they different?

SECTION 16.3 CLASS ACTIVITY 16-K

Critique Probability Reasoning about Compound Events

CCSS CCSS SMP3, 7.SP.8

1. Simone has been flipping a coin and has just flipped 5 heads in a row. She says that because she has just gotten so many heads, she is more likely to get tails than heads the next time she flips. Is Simone correct? What is the probability that her next flip will be a tail? Does the answer depend on what the previous flips were?

2. Let's say you flip 2 coins simultaneously. There are 3 possible outcomes: Both are heads, both are tails, or one is heads and the other is tails. Does this mean that the probability of getting one head and one tail is $\frac{1}{3}$?

SECTION 16.3 CLASS ACTIVITY 16-L

Expected Earnings from the Fall Festival

CCSS CCSS SMP1, 7.SP.7a

Ms. Wilkins is planning a game for her school's fall festival. She will put 2 red, 3 yellow, and 10 green plastic bears in an opaque bag. (The bears are identical except for their color.) To play the game, a contestant will pick 2 bears from the bag, one at a time, without putting the first bear back before picking the second bear. Contestants will not be able to see into the bag, so their choices are random. To win a prize, the contestant must pick a green bear first and then a red bear. The school is expecting about 300 people to play the game. Each person will pay 50 cents to play the game. Winners receive a prize that costs the school $2.

1. How many prizes should Ms. Wilkins expect to give out? Explain.

2. How much money (net) should the school expect to make from Ms. Wilkins's game? Explain.

SECTION 16.4 **CLASS ACTIVITY 16-M** 🍎

Using the Meaning of Fraction Multiplication to Calculate a Probability

CCSS CCSS SMP4, SMP7

Materials If you will do part 2 (which is optional), you will need a pencil and a paper clip to make a spinner as follows: Put the pencil through the paper clip, and put the point of the pencil on the center of the circle. The paper clip will now be able to spin freely around the circle.

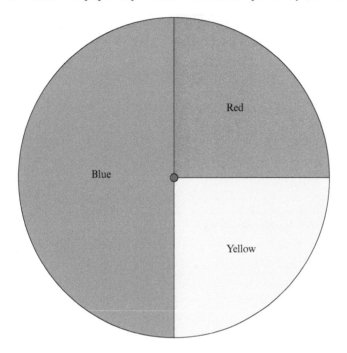

To win a game, Jill needs to spin a blue followed by a red in her next 2 spins.

1. What do you think Jill's probability of winning is? (Make a guess.)

2. If the materials are available, carry out the experiment of spinning the spinner twice in a row 20 times. (In other words, spin the spinner 40 times, but each experiment consists of 2 spins.) Out of those 20 times, how often does Jill win? What fraction of 20 does this represent? Is this close to your guess in part 1?

3. To calculate Jill's probability of winning theoretically, imagine that Jill carries out the experiment of spinning the spinner twice in a row many times, and answer the following questions. You may also want to shade the rectangle below to show the reasoning.

In the ideal, in what fraction of the double spins should the first spin be blue? _____

In the ideal, in what fraction of those double spins when the first spin is blue should the second spin be red? _____

In the ideal, in what fraction of the double spins should Jill spin a blue first and then a red?

_____ of _____ = _____ • _____ = _____

Therefore, what is Jill's probability of winning? Why did it make sense to multiply fractions to determine this probability? Compare the probability with parts 1 and 2.

A rectangle representing many double spins.

4. To win a game, Ronaldo needs to spin a yellow followed by a blue on his next two spins. What is Ronaldo's probability of winning? Explain how to determine this probability with fraction multiplication by reasoning about what would happen in the ideal for a large number of double spins, as in part 3.

SECTION 16.4 | **CLASS ACTIVITY 16-N**

Using Fraction Multiplication and Addition to Calculate a Probability

CCSS CCSS SMP4, SMP7

Materials If you will do part 2 (which is optional), you will need a paper clip, an opaque bag, and blue, red, and green tiles.

A game consists of spinning a spinner like the one below and then picking a tile from a bag containing 1 blue tile, 3 red tiles, and 1 green tile. (All tiles are identical except for color, and the person picking a tile cannot see into the bag, so the choice of a tile is random.) To win the game, a contestant must pick the same color tile that the spinner landed on. So a contestant wins when either a blue spin is followed by a blue tile or a red spin is followed by a red tile.

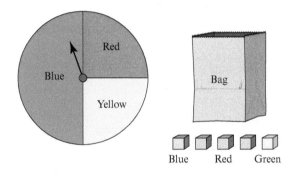

1. Make a guess: What do you think the probability of winning the game is?

2. If the materials are available, play the game a number of times. Record the number of times you play the game (each game consists of both a spin *and* a pick from the bag), and record the number of times you win. What fraction of the time did you win? How does this compare with your guess in part 1?

3. To calculate the (theoretical) probability of winning the game, imagine playing the game many times, and answer the questions below. You might also like to shade the rectangle below to show the reasoning.

A rectangle representing playing the game many times.

a. In the ideal, what fraction of the time should the spin be blue? _____

In the ideal, what fraction of those times when the spin is blue should the tile that is chosen be blue? _____

Therefore, in the ideal, what fraction of the time is the spin blue and the tile blue?

_____ of _____ = _____ · _____ = _____

b. In the ideal, what fraction of the time should the spin be red? _____

In the ideal, what fraction of those times when the spin is red should the tile that is chosen be red? _____

Therefore, in the ideal, what fraction of the time is the spin red and the tile red?

_____ of _____ = _____ · _____ = _____

c. In the ideal, what fraction of the time should you win the game, and therefore, what is the probability of winning the game? Explain why you can calculate this answer by multiplying and adding fractions. Compare your answer with parts 1 and 2.

SECTION 16.4   CLASS ACTIVITY 16-O

Calculating Probabilities with Fraction Arithmetic

CCSS   CCSS SMP4, SMP7

For each of the following two-stage experiments, imagine running the experiment many times, and view the probability of an event as the fraction of times the event should occur in the ideal, over the long run. Use the meanings of fraction multiplication and addition to explain why you can calculate the probabilities with fraction arithmetic.

1. A bag contains 5 green marbles and 2 yellow marbles. To run the experiment, you reach into the bag and randomly pick a marble, put it back in the bag, and randomly pick a second marble.
   a. What is the probability of picking two yellow marbles?

   b. What is the probability of picking two marbles of the same color?

2. A bag contains 5 green marbles and 2 yellow marbles. To run the experiment, you reach into the bag and randomly pick a marble, *leave it out of the bag*, and randomly pick a second marble.
   a. What is the probability of picking two yellow marbles?

   b. What is the probability of picking two marbles of the same color?

DOWNLOADS: bit.ly/2SWWFUX

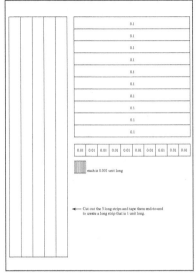

Download 1-1

Download 1-2

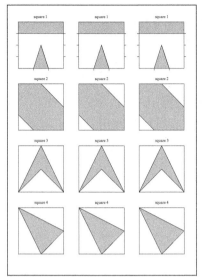

Download 3-1

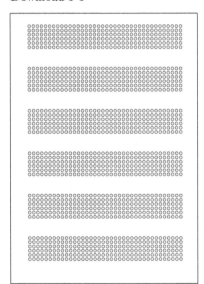

Download 4-1

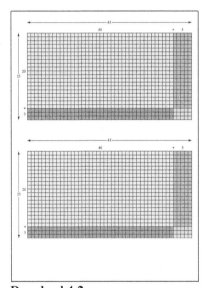

Download 4-2

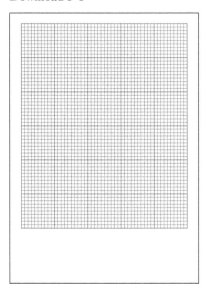

Download 4-3

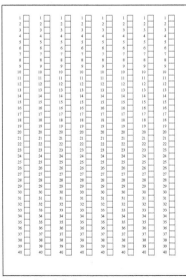

Download 8-1

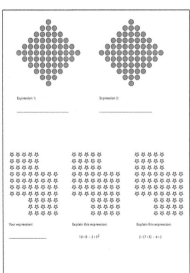

Download 9-1

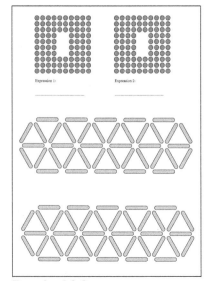

Download 9-2

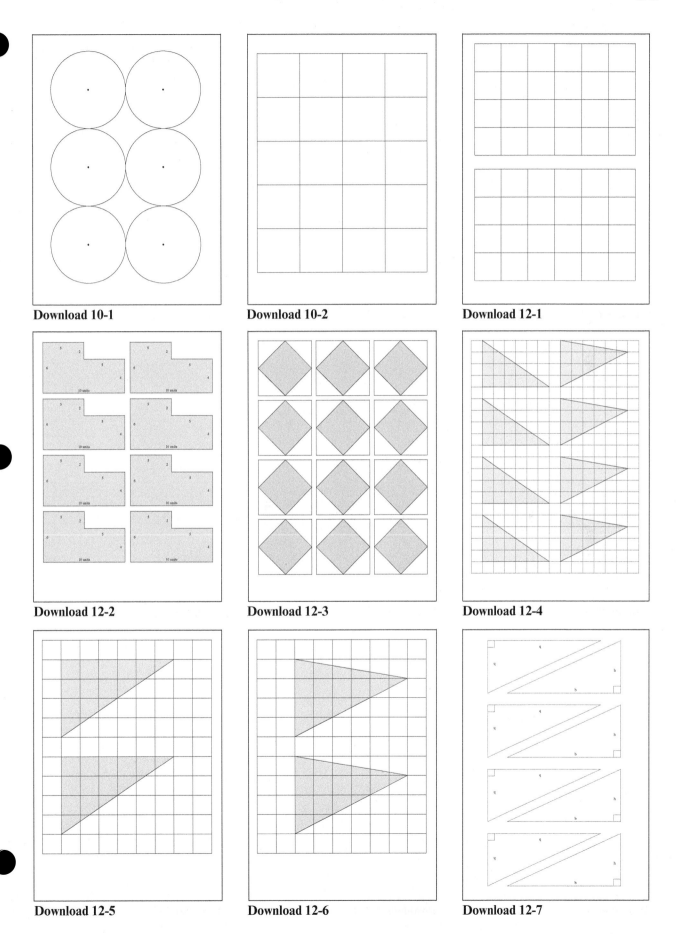

Download 10-1

Download 10-2

Download 12-1

Download 12-2

Download 12-3

Download 12-4

Download 12-5

Download 12-6

Download 12-7

D-3

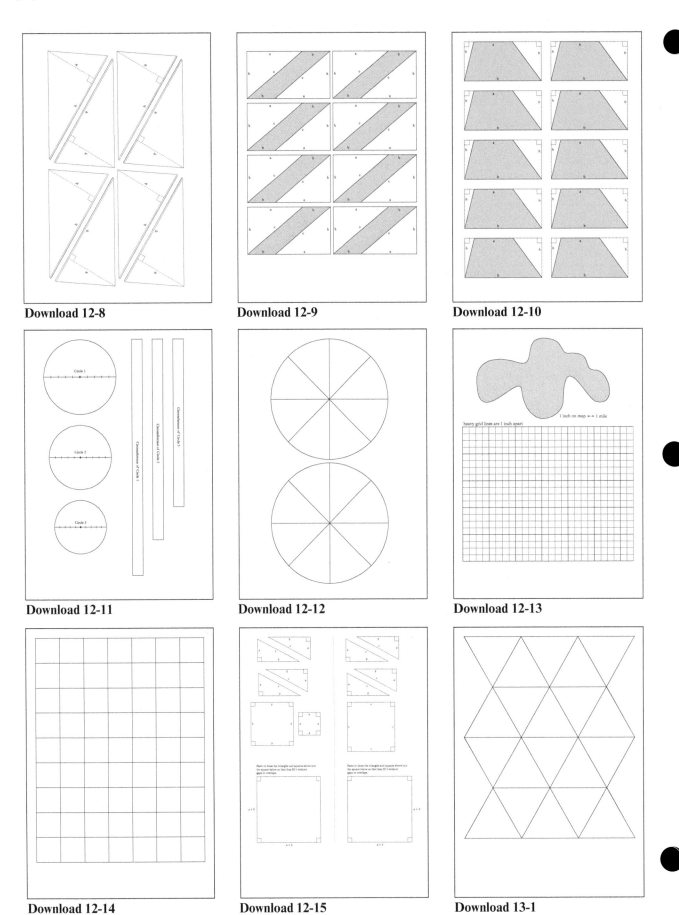

Download 12-8

Download 12-9

Download 12-10

Download 12-11

Download 12-12

Download 12-13

Download 12-14

Download 12-15

Download 13-1

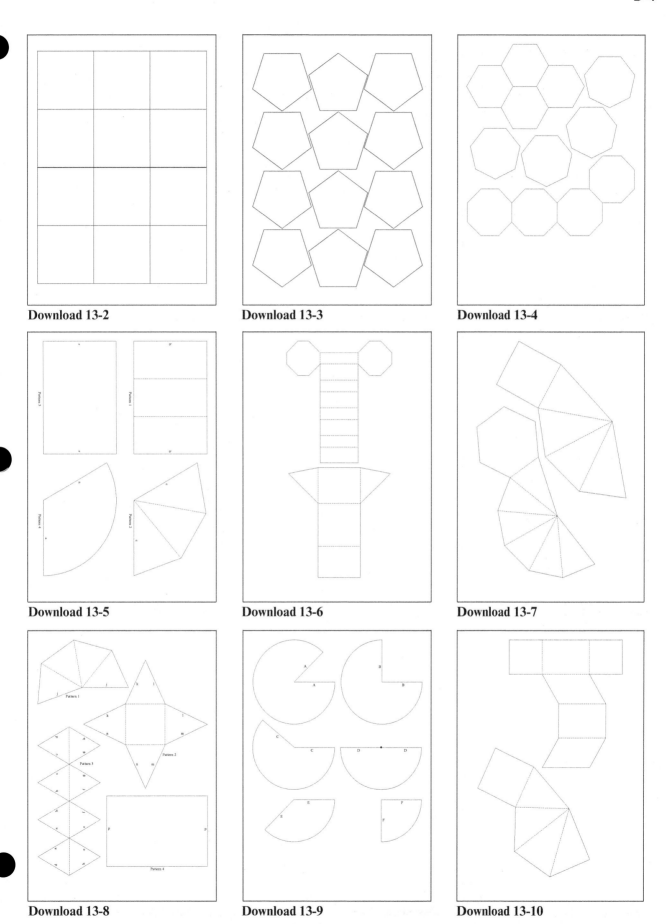

Download 13-2

Download 13-3

Download 13-4

Download 13-5

Download 13-6

Download 13-7

Download 13-8

Download 13-9

Download 13-10

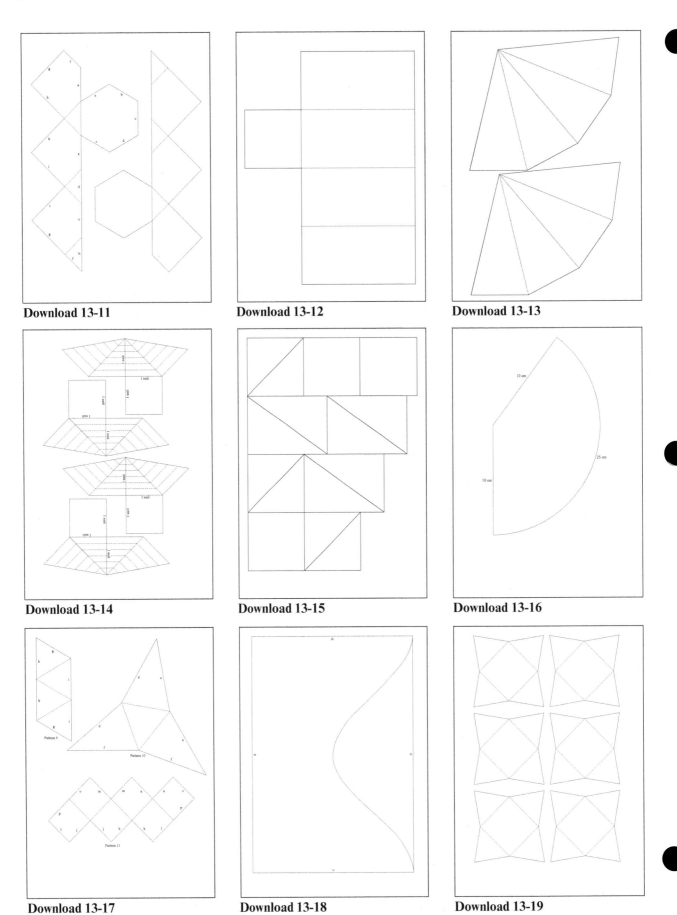

Download 13-11

Download 13-12

Download 13-13

Download 13-14

Download 13-15

Download 13-16

Download 13-17

Download 13-18

Download 13-19

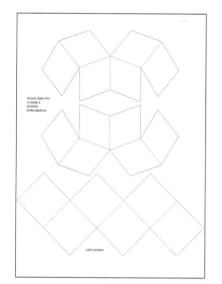

Download 13-20

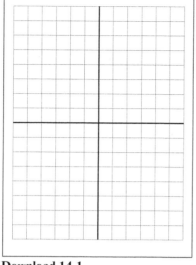

Download 14-1

Download 14-2

Download 14-3

Download 14-4

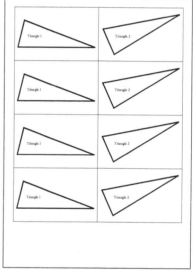

Download 14-5

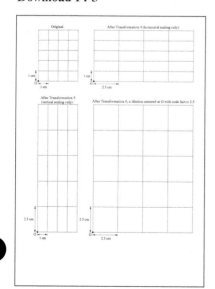

Download 14-6

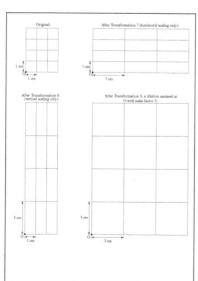

Download 14-7

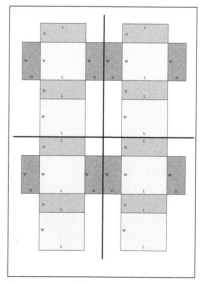

Download 14-8

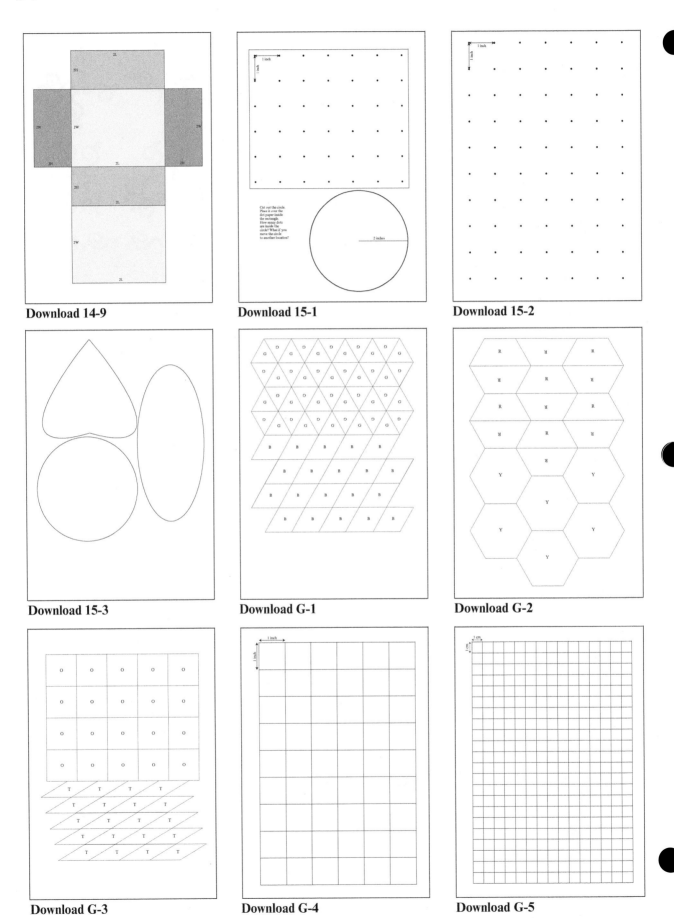

Download 14-9

Download 15-1

Download 15-2

Download 15-3

Download G-1

Download G-2

Download G-3

Download G-4

Download G-5

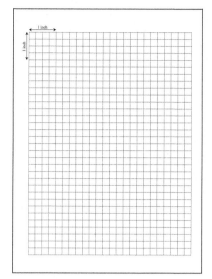

Download G-6

BIBLIOGRAPHY & INDEX

In this edition the Bibliography and Index for the Class Activities are combined with those for the textbook itself, so please consult the back of the Student Edition or the Instructor's Edition for these references. Note that in the Bibliography and the Index, the page references for all Class Activity pages are differentiated with the prefix "CA."